"I started to read this book at six in the morning and did not move out of my chair for the rest of the day. I did not lose a day; I gained a whole world. I invite you to share this extraordinary tale of magic beings who touch us in ways we can only begin to imagine."

—Jeffrey Moussaieff Masson, author of *The Emperor's Embrace* and *When Elephants Weep*

"Living on the edge of her senses, writing down what she knows to be true, Montgomery is a modern miracle: bawdy, brave, inventive, prophetic, hell-bent on loving this planet. . . . She numbs us with the searing beauty of the Amazon world and then numbs us again with proof of our own greed."

—Beth Kephart, *Book* magazine

"The dolphins look you right in the eye, as does the author. Her dealings with all the denizens of the rain forest . . . are searching and personal."

—*The New Yorker*

"Surely one of the most brilliant books of our time. . . . Montgomery weaves zoology with myth, natural history with poetry, anthropology with the supernatural, and the result is perfection, a picture not only of animal life but also of human life in the Amazon Basin. Montgomery has found a new and very perceptive way to present the natural world."

—Elizabeth Marshall Thomas

"Here is Montgomery at her best, beckoning us to follow her in search of one of our planet's most mysterious creatures, and introducing us to a wonderful cast of characters along the way. . . . No one is better than Sy at making you fall in love with nature all over again!"

—Mark Plotkin, ethnobotanist and author of *Tales of a Shaman's Apprentice*

"It's a breathtaking book. This account of a naturalist's experiences in the Amazon turns its own pages, drawing the reader deep into the world of pink river dolphins. . . . Exhilarating, vivid, and often horrifying, this is a serious report on the real and mythical life of an enchanting Amazon species and the sumptuous flooded rain forest."

—Katy Payne, discoverer of the songs of humpbacked whales and author of *Silent Thunder*

"Sy Montgomery brings an adventurer's daring, an artist's sensibilities, and an ethicist's wisdom to field biology. She also, luckily for us, writes beautifully."

—Sue Hubbell, author of *A Country Year, Broadsides from the Other Orders,* and *Waiting for Aphrodite*

"[Sy Montgomery] recounts her adventure and observations with the lyricism and penetrating insights of a poet as well as the logic and factual accuracy of a scientist. . . . Mesmerizing accounts of her daring feats are linked to hard-hitting disclosures of the cruel human history of this bewildering and beautiful realm, but Montgomery's most impressive accomplishment is her illumination of the overlay of story and science."

—*Booklist,* starred review

"Combining a journalist's cool objectivity with a dolphin lover's almost mythical ecological consciousness, Montgomery luxuriates in the myths and legends . . . but also aptly reports the scientific facts. . . . Her rhapsodic book winsomely blends travel, reportage, adventure and natural history."

—*Publishers Weekly*

"She weaves natural history, anthropology, myth and the supernatural in a riveting and often horrifying tale of plants, animals and humans whose lives are inexplicably tied to the river basin. . . . All told, it's a very lyrical and literary paddle."

—Susan Dworski, *Escape*

"Part natural-history exploration, part travel memoir, the book looks to biologists, shamans and local storytellers to reveal a relationship between the dolphin and the river people that is sensuous and powerful. . . . There is Montgomery herself, a writer at the peak of her craft, who places no limits on her quest to fathom the magic and the science of this luminous river."

—Stephen J. Lyons, *New Age*

"Lyrical, erudite, and entertaining. . . . As if these creatures weren't wonderful enough to read about, there are also intrepid lady explorers and scientists; handsome, knowing boatmen; . . . jungle ritual; and even [an] aqua-terrorist."

—Judith Stone, *Mirabella*

"Montgomery is a fine writer, at her best in describing the almost overwhelming beauty and variety of this richest of earth's habitats. . . . [This book] will and should be widely read and argued over."

—Bill McKibben, *The Boston Globe*

"This book will claim you. . . . Heady stuff. . . . Her vivid descriptions of the Amazonian forest . . . make this a recommended choice for all natural library collections."

—*Library Journal*

"Montgomery recounts in vivid and imaginative prose the fascinating story of her quest to track and study these enigmatic creatures along thousands of miles of the Amazon River. One woman's magical search into a watery world of wonder and mystery."

—June Sawyers, *Chicago Tribune*

"[*Journey of the Pink Dolphins*] is like reading Gabriel Garcia Marquez—with natural history!"

—Joni Praded, director and editor, *Animals* magazine

OTHER BOOKS BY SY MONTGOMERY

Walking with the Great Apes: Jane Goodall,
Dian Fossey, Birute Galdikas

Spell of the Tiger: The Man-Eaters of Sundarbans

Nature's Everyday Mysteries

Seasons of the Wild

FOR CHILDREN

The Snake Scientist

JOURNEY
of the
PINK
DOLPHINS
An Amazon Quest

Sy Montgomery

A TOUCHSTONE BOOK
PUBLISHED BY SIMON & SCHUSTER
NEW YORK LONDON TORONTO SYDNEY SINGAPORE

TOUCHSTONE
Rockefeller Center
1230 Avenue of the Americas
New York, NY 10020

Designed by Jeanette Olender
Manufactured in the United States of America

3   5   7   9   10   8   6   4   2

The Library of Congress has cataloged the Simon & Schuster edition as follows:
Montgomery, Sy.
Journey of the pink dolphins : an Amazon quest / Sy Montgomery.
p.   cm.
1. Inia geoffrensis—Amazon River Region.
2. Amazon River Region—Description and travel.   I. Title.
QL737.C436 M66 2000
599.53'8—dc21   99-045840
ISBN 0-684-84558-X
0-7432-0026-8 (Pbk)

In memory of my father

Brigadier General A. J. Montgomery

CONTENTS

Map: The Author's Travels Through the Amazon, 12

A Woman Rain, 15

Manaus: The Curtain Rises, 21
Map: Meeting of the Waters, 23
The Meeting of the Waters, 39
Unfathomable Fragments, 53

Desire, 69

Iquitos, 75
Map: Iquitos, 77
Life in the Rain Forest, 96
Death in the Rain Forest, 113

Breath, 129

Vine of the Soul, 133
A Fortress of Orchards, 153
Map: Tamshiyacu-Tahuayo Community Reserve, 155
Time Travel, 175

Drowning, 201

Mamirauá: Calf of the Manatee, 205
Map: Mamirauá Reserve, 207
The Waters Open, 225

The Moon's Tears, 253

Burning, 257
Map: Tapajós and Arapiuns Rivers, 259
Dance of the Dolphin, 275

Selected Bibliography, 307

To Conserve the Pink River Dolphin and Its Habitat and to
Visit the Dolphins' World, 313

Acknowledgments, 315

Journey of the Pink Dolphins

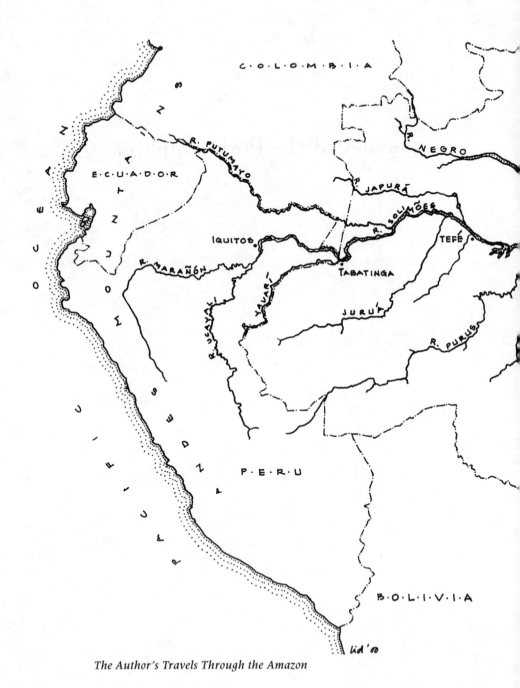

The Author's Travels Through the Amazon

ATLANTIC

OCEAN

MANAUS

R. AMAZON
SANTARÉM

R. MADEIRA

R. TAPAJÓS

R. XINGU

R. TOCANTINS

R. ARAGUAIA

R. TOCANTINS

B·R·A·Z·I·L

AREA OF
DETAIL

SOUTH
AMERICA

NORTH

0   100   200   300   400 MILES.

A Woman Rain

THE DAYS ARE FULL OF WATER. THE WET SEASON HAS DROWNED the village soccer fields and banana groves and manioc gardens, even flooded some of the less carefully placed stilt houses along the river. Young saplings are submerged completely, and fish fly like birds through their branches. Huge muscular trees stand like people up to their torsos in water; epiphytic orchids and the tree-hollow nests of parrots and bamboo rats are at eye level when you stand in your canoe. On the wide branches, tank bromeliads, plants related to pineapples with spiked, succulent leaves, are themselves tiny lakes. The leaves mesh to form an overflowing bowl. Some five hundred different species—centipedes, scorpions, tree frogs, ants, spiders, mosquitoes, salamanders, lizards—have been recorded living in a bromeliad's bowl, a miniworld of rainwater.

Every day there is an extravagant, transforming rain. Sometimes a storm cracks the sky with lightning, crashes branches, rips animals from the trees, and then is gone. This is a man rain, Moises says. But a rain that pours itself out for hours, sobbing and heaving—that, he explains, is a woman rain, "because a woman can cry all day."

As the days are full of water, the nights are full of sound. It is as if all that is revealed by daylight—trees veiled like brides in lianas, leaves the size of canoe paddles, the nests of birds, ants, and termites hanging like tumors, like goiters, like breasts—is transformed, by the dark, into sound. From tree and ground and water, voices rise like little bells and tiny flutes: trills, screams, warnings, laments. They speak in the language of water. In the wet season, the sun grebe's nighttime call sounds like water falling into a tin cup. A ladder-tailed nightjar cries out the name of the river: "Too-WHY-you! Too-WHY-you!" Jewel-eyed tree frogs sing in voices like bubbles rising up from the water, breaking its surface like breath.

The river is the looking glass into another world. By day, the water is a perfect mirror of trees and sky—and yet its glassy surface moves so quickly that if you enter the water without a lifejacket, the current will sweep you under. The river people speak of the Encante, an enchanted city beneath the water, ruled by beings they call Encantados. Those who visit never want to leave, because everything is more beautiful there.

At night, even the stars seem brighter in the water than in the sky. The constellations shine above, their starry reflections below, and from the trees, the glowing eyes of wolf spiders, tree boas, tree frogs. In your canoe, you feel like you are traveling through the timeless starscape of space.

But if you stop and wait, the Encantados will come. At first you may feel a sizzle of bubbles rising beneath the craft, an effusion of pearls cast up from below like a net of enchantment. If the night is moonless, you will only know their breath. But if the moon is full, you may see a form rising from the water, gathering into the shape of a dolphin. Inches from your canoe, a face may break the surface—a face at once other-worldly and eerily familiar. The forehead is clearly defined, like a person's. The long beak sticks out like a nose. The skin is delicate, like ours. Sometimes it is grayish, or white—and sometimes dazzlingly, impossibly *pink*. The creature turns its neck and looks at you, and opening the top of its head, gasps, "Chaaahhhhh!"

In Brazil, they call this dolphin "boto." They say the boto can turn into a person, that it shows up at festas to seduce men and women. They say you must be careful, or it will take you away forever to the

The flooded forest: a looking-glass world.

Encante, the enchanted city beneath the water. In Peru, they call the creature "bufeo colorado"—the ruddy dolphin. Shamans say its very breath has power, and that the sound it utters when it gasps can send poisoned darts flying, as from a blowgun. Scientists call it by the species name: *Inia geoffrensis*. They say it represents an ancient lineage of toothed whales—a freshwater dolphin, caught in a Miocene time warp since the era when alligators bigger than *Tyrannosaurus rex* lurked in the shallows, when flightless, carnivorous terror birds six feet tall seized prey with three-fingered, saber-tipped hands.

Each person who encounters an Encantado is touched by its en-

chantment. Each comes away from the encounter speaking a different truth, informed by dreams and ghosts and the hot, whispered breath of rain on the river. For here in the Amazon, where unfathomable tragedies collide with unquenchable desires, the most preposterous of impossibilities come true.

MANAUS: THE CURTAIN RISES

A thousand miles up the Amazon River, in the middle of the world's largest rain forest, an opera house rises from the jungle.

By steamship, the building crossed the ocean piece by piece. Its iron framework came from Glasgow, marble from Verona and Carrara, crystal from Venice, cedar from Lebanon, silk from China. One hundred crates of ornately carved furniture, upholstered in velvet, were imported from London. Only the wood of the parquet floor—12,000 pieces of oak, brazilwood, and jacaranda—originated in Brazil. Even this was hand-worked in Europe before Portuguese craftsmen, themselves ferried across the ocean, mounted each piece in place without the use of nails or glue.

The walls inside were painted by Italy's leading creator of sacred murals, Dominico de Angelis, in the style of the grand cathedrals of Europe. Ceiling frescoes decorated with cherubs and angels depict the four great arts: to stage right, Dance; to stage left, Music; at back, Tragedy; and in front, the supreme art, Opera, which combines all three. A mosaic of 36,000 vitrified ceramic tiles from Alsace-Lorraine crowns the theater's cupola in glittering blue, gold, and green. The dome rests atop a neoclassical confection of twisting, balustraded stairs and columned porticoes, bordered in white, like icing piped on a wedding cake. In this city amid the jungle, the building's pastel hue, refined and delicate, seems as lurid as an orchid. The facade is pink—the color of the dolphins who inhabit the waters that brought this city its impossible wealth.

It took more than fifteen years and $2 million to construct the Teatro Amazonas. Inaugurated in 1896, it was lauded as the most beautiful opera house in the world. It is said it was built to attract Enrico Caruso. But he never came. An epidemic of yellow fever killed 16 members of the Italian opera company performing at the time the great tenor was invited. Some 300 of the city's citizens died of malaria each year. Newspapers advertised potions to counteract the snakebite from bushmasters. Like a Wild West frontier town, the city had to enact an ordinance (largely ignored) forbidding the firing of guns and arrows on the streets.

Yet in the opera house's harp-shaped theater, audiences of 1,600 gathered, dressed in diamonds and silks. In turn-of-the-century Manaus, diamonds were the unofficial currency of the bank of rubber. At the height of the motoring world's demand for tires, rubber, the milky lifeblood of the rain forest's seringueira tree, was ferried here from the Amazon's thousand tributaries to drench Manaus in opulence. The citizens of Manaus became the highest per capita consumers of diamonds in the world. Some women set their teeth with them, diamonds glinting behind the flutter of black lace fans. A waitress serving a sandwich to a lunchtime customer might receive a diamond as a tip. A top prostitute could expect a diamond necklace as payment. Prices in Manaus were four times those of New York, and druggists could charge two British pounds for a shilling's worth of quinine. Yet rubber barons slaked their horses' thirst with French champagne. Bathroom faucets were set in solid gold. Housewives sent the linens to Portugal to be laundered.

But some stains would never come clean. To gather and process the

Meeting of the Waters

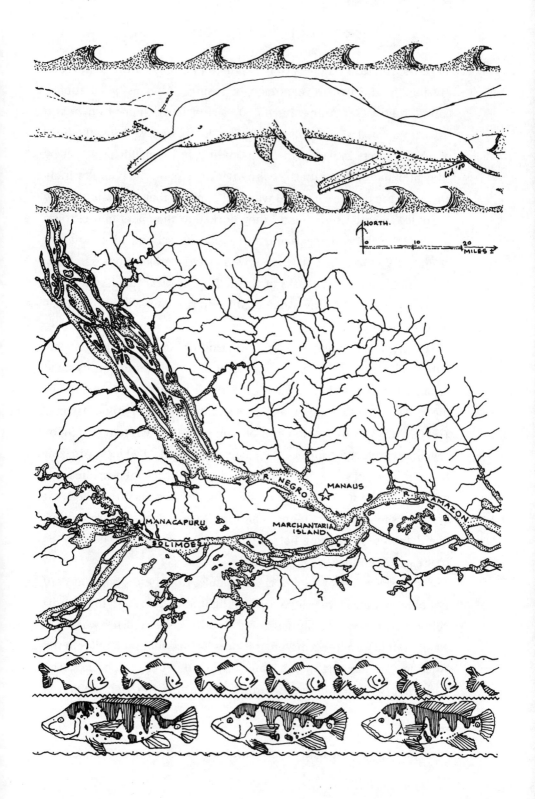

latex, the Amazon's Indians were captured in chains, tortured into sub-
mission. In the twelve-year reign of a single Manaus-based rubber
baron, Júlio César Arana, his 4,000 tons of latex shipped down the
Amazon fetched $7.5 million on the London market and cost the lives
of 30,000 forest Indians. "In truth," writes historian Richard Collier,
"the latex barons built their wealthy wicked city on the bones of Indians—and the people's frenzied way of life hinted that they knew it."

So, like penitents to church, they flocked to the opera house, to sit
beneath the arts and angels and surround themselves with the names
of Europe's greatest artists. On the twenty-two marble columns in front
of the noble boxes are masks of Greek Tragedy, bearing plaster scrolls
inscribed Goethe, Rossini, Molière, Shakespeare, Mozart, Wagner, Bee-
thoven, Lessing, Verdi . . .

The masks of Tragedy faced the Amazon of Europe's longing imagination. The centerpiece of the opera house is the painting on its stage
drop curtain, *The Meeting of the Waters*. It was created by a Brazilian living in Paris, the Comédie Française's scenic artist, Crispim do Amaral.
The curtain rises to the dome in one fluid motion, without folding or
rolling, to protect the image. The pale-skinned, naked goddess Amazon
floats on a bed of gossamer, supine and pliant. She leans back, one knee
raised, as if about to open her thighs to the two bearded river gods on
her opposite sides, the Solimões and the Negro. They rise around her
like dolphins. They bring her garlands of flowers. Their waters are as
blue as the Danube.

The real Meeting of the Waters is six miles from here. The waters are
not blue, but an extraordinary mix of light and dark. Here the cream-
colored, white-water Solimões and the coffee-colored, black-water Rio
Negro join to form the Amazon as it runs its final thousand miles to the
Atlantic. The Meeting of the Waters is a confluence of opposites. The
dark waters of the Negro are born in the ancient Brazilian highlands,

The Meeting of the Waters, where dark and light waters flow side by side.

whose sediments leached away millions of years ago. Because these sandy soils are too poor to break down organic chemicals, the Negro is laced with acids and stained with tannins. Its waters are nearly sterile. The muddy water of the Solimões owes its light color to huge quantities of nutrient-rich silt gathered from headwaters in the geologically youthful Andes. It teems with piranhas and electric eels, fish with bony tongues and bulging eyes. Pink river dolphins come to hunt here, at the Meeting of the Waters, where the fish, confused by the sudden collision of waters, make easy prey.

For nearly four miles, the two rivers, because of their different densities, meet but do not mix. Side by side they flow, little fingers of dark and light clutching at one another like lovers who cannot marry. The Amazon is born of such a union: a confluence of separate histories, of opposite identities, a meeting of beauty and cruelty, desperation and passion, life and death.

Tomorrow night, I will sleep at the Meeting of the Waters, on the floor of a wooden house that floats on logs of assacú, tethered to a

three-boled taro matue tree. But at the moment, as mosquitoes chew my ankles, I gaze over a red velvet rail, eye level with a French chandelier of gold and crystal, surrounded by Molière and Mozart, flanked by Dance and Tragedy.

The drop curtain has been raised, showing only the bottom of the painting. The waters of the Amazon seem to pour over its painted gilt frame, out into the audience. The orchestra is tuning up. The kettle drums sound like thunder.

Thunder woke me my first night in Brazil. Its explosion swept through my skeleton. Lightning flooded the room, so bright I could see it through closed eyelids. The rain pounded at the walls and roof like some frantic jungle demon. I could not get up to look out the window; the rain held me motionless with its force.

Dianne Taylor-Snow lay in the room's other bed, the tip of her cigarette glowing orange. "Toto, I don't think we're in Kansas anymore," came her voice in the darkness.

As usual, she'd been lying awake for hours. Insomnia is but one of Dianne's talents—she can also pee off the back of a moving boat and swear in Indonesian—but her sleeplessness makes her a uniquely stimulating traveling companion. Five years ago, on assignment in Bangladesh, she woke me up at three in the morning to point out a spider in the room—one of her few fears. The spider, about the size of a fist, was sitting on the mirror. But when she shined her flashlight on it, its hugely magnified shadow covered the opposite wall, its head the size of a melon, its hairy legs stretching for twelve feet.

In Bangladesh, looking for tigers, we had seen our first river dolphins. By boat, we were exploring the muddy waters of Sundarbans, the greatest mangrove swamp on earth, the home of the world's largest population of tigers. Yet Sundarbans's tangled mangroves and thick brown rivers, it seemed, connived to conceal everything from us.

One day the muddy waters parted for an instant. Breaking the surface, the curves of three large pink-gray forms rose and rolled, like the play of sunset on water—yet it was midday, the sun bright. We stared after them. Again the forms rose and sank, smooth as silk. Finally, I realized what we had seen: dolphins!

For me, the sight of dolphins anywhere has always carried the shock of recognition—like seeing my own reflection in the water. I had seen wild dolphins in U.S., U.K., and New Zealand waters, and of course in many aquaria, and yet they always surprise. They are shaped more like fish than mammals, and they inhabit what to us terrestrial creatures is a foreign universe—but still, both species seem to know we are in many ways alike. John Lilly, a medical researcher whose studies of how people think and communicate led him to study communication between people and dolphins, calls dolphins "Humans of the Sea." To see a dolphin emerge from the water feels to me like glimpsing a lost twin.

That dolphins could inhabit such muddy waters as Sundarbans's seemed impossible. But then I remembered having read about them: early explorers had been astonished, too, by these river-dwelling whales, and judged (incorrectly) that in waters so opaque, the dolphins must be blind. Later researchers found that the dolphins can see, but navigate mainly by a sonar system that even by dolphin standards is almost unimaginably refined.

After only seconds, they sank from sight like a dream. Yet for a long time afterward, their image glowed in my mind. It was as if the river had opened, for just a moment, and revealed to me some promise.

Three times after that, I returned to Sundarbans. I saw dolphins every time. In each instance, the glimpses were fleeting, unexpected, revelatory—and then they were gone. I never saw a face, or even a flipper—just the top of a head or the low curve of a dorsal fin. I never was able to find much on them in the scientific literature. Almost nothing is known about them. There are, in fact, five species of river dolphins around the world, and none of them is well understood.

Yet the image of the river dolphins stayed with me. Back home in New Hampshire, pink dolphins swam seductively through my dreams.

Years later, at a marine mammals conference, I met a man who told me why: pink dolphins capture souls.

Half a world away from Sundarbans, in the Amazon, he told me, lives a different species of pink dolphin, *Inia geoffrensis*—*inia* is the Guarayo Indian word for "dolphin"—named for Geoffrey St. Hilaire, who plundered the first zoological specimens from Portugal for Napoleon Bonaparte. Unlike the Sundarbans dolphins, they are bold and abundant. But they are equally mysterious.

Photographs of the pink river dolphins looked eerily familiar. They resembled no other dolphin I had ever seen, with their melonlike foreheads and long tubular snouts. Yet they reminded me of something. Then I realized: It's *us* they look like, but in another form. These pink dolphins look like a fetal human, a person in a watery beginning.

A number of researchers at the conference were trying to study them, but with limited success. Few scientists claimed to know the number of dolphins in their study area, or to understand the social structure of groups, or to know whether they migrate or hold territories. They did not even know for sure why these dolphins are pink: some said only old animals are pink and that the young ones are always gray; others said the dolphins flush pink with excitement. In fact, even after many years of study, most researchers confess they cannot even recognize individuals on sight.

To local river people, this confusion is no surprise, my informant told me. The people say these dolphins are shape-shifters. In the guise of human desire, they can claim your soul, they carry you away, and they take you to the Encante, an enchanted world beneath the river.

In fact, the pink river dolphins do inhabit an enchanted world: the Amazon. I had always longed to go there. In the western imagination, it has invoked an El Dorado, a Last Frontier, a Green Hell, a mythical race of woman warriors, an earthly Paradise, a Paradise Lost. To the people whose villages ring its waters, the Amazon is the source of renewal and destruction, of powers and inspirations. The scientists tell us that the Amazon holds one half of the world's river water and that its leafy basin supplies a tenth of the world's oxygen; our connection to it is as close as breath. And yet, like the dolphins, the Amazon remains a great mystery, from which we seek to satisfy dizzying desires, to extract livelihoods and longings, and onto which we project our deepest fears, our darkest appetites.

I knew, at that conference, what I must do next. I would follow the dolphins. Dianne eagerly volunteered to come with me.

And now, at the start of our expedition, already they h d led us to this impossible city: this Paris in the jungle, where the wa er that becomes the river fills even the sky. At the Meeting of the Waters, we were told, we would find dolphins. We didn't know where they would lead us.

But we knew, from that first soaking storm in Brazil, that wherever it was, it would be wet.

"My poncho smells like cheese," I said to Dianne, as a second wave of thunder rolled through our bones. I was hoping that another poncho numbered among the contents of her suitcases; she always traveled well provisioned. Once, on a trip to Cameroon, when she and her companion had to stay in a tiny hotel with filthy linens, she pulled from her luggage, to her companion's amazement, a set of satin pillowcases. Another time, after she'd picked me up at the L.A. airport on a book tour, I was rummaging in the glove compartment for a map and found instead a Colt .380 semiautomatic pistol. "There is a handgun in your glove compartment," I reported in alarm, as if it had somehow appeared there by mistake. "Oh, that's just my friend Fluffy," she said. Her familiar tone made clear she knew well how to use it.

On this trip, I noted thankfully, Fluffy had stayed at home. As usual, Dianne had spread the contents of her luggage all over the room, like some sumptuous buffet: the miniature hair dryer with its new adapter; a hammock that balls up to the size of a grapefruit; the inflatable travel pillow (from her days as an airline stewardess); the four clear plastic, zippered cases of lip gloss and eyelash curlers and eyebrow pencils and shampoos (from her days as a fashion model); baby bottles and nipples and Esbilac in case we encountered orphaned baby animals (from her days as an orangutan rehabilitator); packs of hermetically sealed portions of raisins, peanuts, and powdered Gatorade; pills to kill intestinal worms; creams to soothe scabies; Ziploc bags of underwear and khaki pants and silk shirts; a heated styling brush; the surgical kit with its scalpels and disposable syringes; the $400 Katadyn water filter that screens out viruses; buck knives and lighters as presents to villagers; a stretchable laundry line; a device to suction snakebite; and clothes for the opera.

But Dianne's mind was on breakfast. "What kind of cheese?" she asked me. "Feta, Gorgonzola . . . ?"

"Cottage," I said, and let the rain drown my consciousness in a water-dream of sleep.

By morning, the storm had dissolved into the washing, ticking, clicking sound of rain in a city, mixing with car horns and police whistles. On the wet streets of Manaus, the asphalt shone like a river. Cars slid by like canoes. The roof of the hotel was leaking, and ragged terry-cloth bath mats sopped up water from the tiled floor of the lobby. The rooftop swimming pool overflowed.

As we planned the chores to prepare for our expedition, from the twelfth-floor restaurant of the Hotel Monaco, we looked down on the Teatro Amazonas's glittering dome, the Victorian Customs House, and the ingenious floating docks—built by British engineers—that rise and fall thirty-two feet a year as the waters of the Rio Negro swell and empty.

The idea to create a "Paris of the Tropics" had been the grandiose vision of one man: Eduardo Gonçalves Ribeiro, a black military engineer who, at the unlikely age of thirty, became the youngest governor of the immense state of Amazônia. "I found a village," he later boasted; "I made of it a modern city." He was a man of small stature, but large appetites—for gold, for women, for fame. Before he even took office, he began laying plans for his favorite project, the opera house. It bears his name in letters tall as a man on the outside wall of its pink facade.

Levying a 20 percent tax on all rubber that left the city, he built streets a hundred feet wide and paved them with cobblestones imported from Portugal, bordered with barbered shade trees from Australia and China. He commissioned the designer of the Eiffel Tower to build a municipal market to look like Paris's Les Halles, and a Palace of Justice that resembled Versailles. Manaus had electricity before London, a telephone system before Rio de Janeiro. While New York and Boston relied on horse-drawn trolleys, Manaus enjoyed the opulence of bottle-green electric streetcars that operated around the clock. Looking down from the rooftop, Dianne and I imagined ladies dressed in Surrah silks, their hair ornamented with the feathers of egrets.

But as we ventured into the city, we found, instead, women in tight Lycra dresses, their hair decorated with yellow plastic barrettes shaped like Tweety Bird. The cartoon character incongruously accents clothes so suggestive that at first, we thought they were streetwalkers' garb. But in the *casa de câmbio* where we changed money, in line at the *supermercado*, on the bus to the Amazon's research institute, we noticed almost everyone dresses like this: halter tops in which each breast seems to be riding in a private hammock, stretch pants so clingy you can see the outline of lace on the underwear beneath. Even fat and wrinkled old ladies, even women who are grossly pregnant, wear short, tight dresses or plunging halter tops or jeans popping at the seams—often with the top button unbuttoned to accommodate the overflow.

Everywhere is a flood of flesh. In the fish market, the big bellies of bare-chested old men spill over the waistbands of their pants like the foaming head on a beer. Manaus revels in the ripeness of flesh and fat—breasts, bellies, buttocks. All these, no matter what shape or age, seem to be greeted with generous approval and enjoyed like public art. The same is true of music. Stores, buses, and restaurants blare samba, rock, and boi-bumbá music, and it spills out into the street, spewing sound like the waters of a public fountain. Even litter is dropped as if this, too, is an act of generosity, performed for the greater good.

There is a feeling of abundance in Manaus: it seems as if the fullness of the wet season has unleashed a torrent of fecundity. Even the trees that line the street are laden with food—cohi, guava, mangos, avocados—and they release their fruits with lush abandon onto the sidewalks. The streets smell of guava and grease and of the meat roasting on every corner, grilled on wooden skewers laid over coals. One little girl is cooking chicken feet this way. She is heartbreakingly beautiful, with the lush, natty hair of an African, the high cheekbones of an Indian, and the green eyes of a Portuguese. She gives us a brilliant smile.

We are in the heart of a country that has been described as "the leading producer of human misery." Every six seconds a Brazilian baby dies of diarrhea; every thirty minutes a Brazilian contracts leprosy and another contracts tuberculosis. There are a million cases a year of malaria, and 10 million of schistosomiasis, a blood fluke that eats through the liver. And yet, in Manaus, where Portuguese sounds like it is spoken through lips numb from kissing, you feel caught up in the sensuous savor of life. In India, they call this rasa: the sweet sap, the juicy life-essence, the core of enjoyment of food, or art, or sex.

In Manaus, there are two great temples that honor the rasa of its people. One is the opera house; the other is the fish market.

Beneath the iron roof and stained glass of the Municipal Market, dark men wearing gold necklaces hold fish up proudly for us to photograph. Young men with sweating chests and tensed muscles carry huge sacks from the docks, green eyes glowing in dark faces like emeralds by firelight. Here, at the largest freshwater fish market in the world, some of the strangest fish on earth are for sale, dredged up from the dolphins' underwater city. Huge slices of silvery pirarucu drape over the counter

An ode to appetite: the Manaus fish market.

like tablecloths. Tucunaré with ruby eyes are arrayed like jewels; their tails glow with eyes, too, like the tail feathers of peacocks. These fish are pursuit predators; once they attack, they do not give up until they have swallowed their prey whole. There are piles of fat, black tambaqui, a sweet-fleshed seed-eater who inhabits flooded forests, and pyramids of piranhas, and careful stacks of predatory catfish, some with black armor, some with long, fleshy whiskers like a Chinese emperor, some with erectile, poison-tipped spines.

Each year some 30,000 to 50,000 tons of river fish, of more than two hundred species, are landed in Manaus. But as the fishing fleet ex-

pands—today nine hundred boats now can hold a ton or more of fish, and a few can hold fifty—the fish they catch get smaller. In the 1970s, I'd been told, you could often find for sale here pirarucu longer than a canoe. The largest fish we see today is less than three feet long. The species has already been fished out of some of the smaller river systems. Tambaqui was once so prevalent that it was fed to prisoners. Twenty years ago, tambaqui accounted for half the total catch. Today most of the tambaqui for sale are juveniles, sold at prices that rival the choicest cuts of beef. That is one reason so many catfish species are for sale: pirarucu, tambaqui, and tucunaré are now too expensive for the average consumer.

But in the fish market, we sense no feeling of dwindling commodities, of time running out. There is only the siren song of the strange and the beautiful, the throb of appetite. Dianne and I are voyeurs at a dazzling show of abundance. Fish with tails striped like tigers, fish with eyes flecked with gold, fish with huge scales rough as emery boards, and fish whose tongues the people use to grate their tuberous manioc to make farina—on the market's aisles of white-tiled tables, each fisherman lays out his catch as lovingly as he would the corpse of a relative. The men rinse the fish regularly with buckets of water that carry away blood and scales in glittering waterfalls to the floor.

In the meat stalls, the carcasses of pigs, cattle, and lambs hang dripping from meat hooks, hooves and heads still on. The men's hands are slippery with blood. Behind the vegetable stalls, smiling men and women chat over mountains of produce: softball-sized avocados that smell like hand cream; parsley with its roots still on, gasping for the soil; yellow melons, shiny fragrant limes, and strange fruits with alluring names like genipapo, acerola, maracujá, pitomba. It is impossible here not to think, at once, of sex and death: the dead fish with their big teeth;

Fish with bony tongues, fish with jeweled eyes at
the Manaus fish market.

the fruit, ripe and overripe; the flecks of blood splashing up at you as butchers sling down slabs of flesh, blood reaching for your blood; and you can feel the eyes of men slide over your body like a tongue licking an ice-cream cone. Life, says the market—choose life! A toothless old woman offers us a taste of maracujá: paper-skinned, seed-filled, its flesh is slippery and bitter, like a mouthful of semen. I swallow the seeds.

The houselights dim. The spotlight rises like a moon, the red velvet curtain parts to a storm of applause. We spend our last night in Manaus at the opening of the first opera to be performed in the Teatro Amazonas in nearly a century: the State Academic Opera and Ballet Theatre of Minsk's performance of Verdi's *La Traviata,* a belated celebration of the opera house's centenary.

   Onstage, a sumptuous party gathers. Women in flounced gowns of

velvet and satin flutter lace fans and flash jeweled tiaras; men strut in tuxedos and white gloves. It is a party at the home of the courtesan Violetta Valery, where the heroine will meet her future paramour, the handsome nobleman Alfredo Germont. But the staged scene is eerily reminiscent of the gathering of the last patrons of the opera to fill this room, ninety years earlier—before anyone imagined that Asian rubber plantations, grown from seeds smuggled from the Amazon, would bring a total collapse of Manaus's rubber boom economy.

In the panic that ensued, women pressed their diamonds into the hands of the cashier at the Booth Steamship Company, to pay for passage to Europe or the United States. In 1912, 140 of Manaus's finest mansions were sold at auction. People fled in such haste that they left their good-byes to friends in the personal columns of the *Jornal do Comercio*. The Paris in America fashion store halved the price of its perfumes, its panther-skin rugs, its Steinway grand pianos. Civil servants went without paychecks. Students rioted. Colleges closed. The stage of the opera house went dark, for the curtain had fallen on the Golden Age of Manaus.

But the ghosts linger. With the violins' pleading, the rising notes of the soprano and tenor's duet summon the lost, sweet dreams: They would have been brave, the women and men who came here to seek their fortunes, to live in the jungle when a mosquito or a glass of water could kill you and not all the silks and brocades and champagne in the world could help. Luxury could never buy safety. They may have sent their linens to be laundered in Portugal, but at home, their children still died of fever in their arms.

Onstage, Violetta's voluptuary heart has been touched by the love of Alfredo. She knows he is not rich; she knows she will bid good-bye to this house and its gay parties if she goes to live with him. Splendid like a bride in her long white gown, she sings, alone, of the transformation her new love has brought. She now sees all the pleasures of her previous life as hollow. The red velvet curtain closes on her solo.

I asked Dianne how she liked the opera so far, the first she had ever seen. "I didn't expect," she said thoughtfully, "that it would be so loud." Then she went outside to smoke a cigarette.

The opera house is lit at night, like some luminous ghost of the past, rising out of the darkness of swallowed memories. The great ethnobotanist Richard Evans Schultes refused to enter the building when he first visited Manaus in 1944. It was built, he told his Brazilian hosts, "with the blood of Indians." He knew the twisted appetites of the overseers whom the rubber barons hired to ensure the flow of cheap latex. On a river named Madre de Dios after the Mother of God, one merchant kept 600 Indian girls as slaves he bred like livestock to produce future laborers. Indians were blindfolded and tethered as rubber traders shot off their genitals for diversion. In Peru, to punish workers who fell short of their quota, one overseer ordered a massacre of Indian children and had them cut up for food for the guard dogs. Another celebrated Easter by personally shooting 150 Chontaduras, Ocainas, and Utiguenes. Those his bullets did not immediately kill were heaped into a pile, soaked with kerosene, and burned alive.

Their ghosts are here, too, in the opera house. The air feels thick with spirits seeking solace. There are ghosts of the Italian musicians and singers who died in epidemics of malaria and yellow fever. It is said that the ghost of the governor, Eduardo Ribeiro, is here as well. He strangled on his own demented desires, and died "in a fit of erotic mania" three months before his beloved opera house would open.

These ghosts have lingered for a hundred years; but this is the first time in ninety they have heard the music of an opera.

In the second act of *La Traviata*, the couple is living at Alfredo's house near Paris, and Violetta has sold her jewels to meet their expenses. But despite her noble gesture, she is still a former courtesan, and an embarrassment to Alfredo's family. In a baritone as commanding as a cannon, Alfredo's father begs her to give up his son; and in notes as yielding as water, Violetta agrees to make the sacrifice, to let Alfredo believe she has left him to return to her life of sin and luxury. She flees to Paris.

Their songs rise, with the warmth of breath, from the stage to the balconies to the ceiling. My attention drifts, with the music, upward, to the angels on the dome. The frescoes of the Four Great Arts were painted with a nineteenth-century technique employed in many cathedrals: it gives the impression that the eyes of the people in the images are always following you. Those who built this place wanted angels to guard them; but instead of guarding, these angels witnessed. And perhaps the people realized this. Perhaps all the music that has played here has been an offering to these witness-angels—music rising like burned incense. Tonight, they seem to watch the opera and its audience with eyes full of knowing.

The opera house stands for everything Europe has tried to make of the Amazon; its attempt to turn the brown waters of the Rio Negro blue as the Danube, and change its white latex to gold. This pink confection erected in the jungle was, perhaps, an attempt to cleanse Europe of its guilt, to transform its lust to beauty.

And this is the transformation played out onstage. In the third act, Violetta, the wealthy courtesan, is now in self-imposed exile, her jewels gone, her health failing from consumption. Her maid brings her a letter—Alfredo's father has told his son the truth of why she left him. Father and son come racing to her side. Alfredo vows to restore her soiled reputation, to take her back to his country home. She tries to rise from her chair to go with him, but she is dying, still in her dressing gown, clad in white gossamer like the angels. Her spirit slips from her body, and in the soprano notes of her song, her soul is released.

Nourished by music, the faded figures overhead, the angels of Music and Opera, Dance and Tragedy, the ghosts of divas and Indians, seem to grow vibrant, to come to life. Weightless as swimming dolphins, they seem to dance in the dark, and beckon us into the jungle.

THE MEETING OF THE WATERS

The morning of our departure for the Meeting of the Waters, only one chore stood between us and the dolphins: the purchase of a frozen chicken.

Unfortunately, "I would like a frozen chicken" was not on my language tapes. So, as I stood at the counter of the *supermercado,* I made up a sentence: *"Queria um frango gelado, por favor."* But I misused a crucial word: *gelado.* As Dianne and our guide sweltered in the truck that was waiting to take us to our boat, I was asking the clerk for chicken ice cream. She looked perplexed. I became increasingly nervous.

To proceed without the frozen chicken was unthinkable. This was the last thing Vera da Silva had told me the night before, in the last of a series of conversations that left me feeling the expedition was falling apart before it ever began.

Vera was supposed to have accompanied us on this trip. The director of the aquatic mammal laboratory at the Amazon research institute, Instituto Nacional de Pesquisas da Amazônia (INPA), Vera has studied pink dolphins longer than any other researcher. She is the only scientist who has tracked them with radiotelemetry. We had first met at the marine mammals conference in the States. There she had generously agreed to take us to the pink dolphins at the study site where she had begun her behavioral work eleven years earlier, a floodplain lake called

Marchantaria along the Rio Solimões, just past the Meeting of the Waters.

I had read and admired Vera's work for months. It had been one of her papers, coauthored with her late husband, Robin Best, a research associate of the Vancouver Aquarium, that had given me the idea of something new I might be able to find out about the animals. "*Inia* would seem an ideal candidate for inclusion in the Convention on Conservation of Migratory Species," they had written in 1986. "Cyclic seasonal migrations due to the annual river floods are almost certain to occur with *Inia*. . . ." The paper had gone on to list a number of loca-

tions where the migrating dolphins surely crossed national boundaries. So I had proposed to try to follow the dolphins on their migration, to see where they go.

On the strength of that proposal, I had secured the money to finance the trip. So when we'd visited Vera at INPA upon our arrival in Manaus, I was especially eager to seek her advice on how to track the dolphins' migration.

"But botos do not migrate," she said.

"But I thought I had read this in your paper."

"Yes," she said, "we thought that at one time, but that was before the telemetry. Now we know they do not migrate."

Oh.

Still, to help us learn about the dolphins, Vera would be indispensably illuminating. And besides, we liked her. She made us immediately at home in her small office, where she had covered the walls and plastered the gray metal file cabinets with posters and postcards of manatees, dolphins, whales, seals, and seabirds. One file drawer bore a 1913 quote from Rebecca West, which must have seemed distressingly current to Vera, working in a field dominated by men in macho Brazil: "I myself have never been able to find out precisely what feminism is: I only know that people call me feminist whenever I express sentiments that differentiate me from a doormat."

Charming and knowledgeable, smooth and supple as a sea creature, in the twenty years she has studied these dolphins she has come, in a way, to resemble them: in the curve of her lips when she smiles, and when her dark eyes dance with humor. Her laughter reminded me of bubbles, like those a boto might release beneath a canoe. "If you stop your boat for several hours, they just come over, touching it, releasing bubbles that are coming to the bottom of the boat," she told us, almost dreamily. "And also you notice when they are coming because they re-

Vera da Silva has studied pink dolphins for more than twenty years.

lease these bubbles as they swim. It's beautiful—a beautiful noise, the air bubbles. And they can be just beside your boat and look at you! It's very amazing. You will see."

I asked about her study technique: What times of day were best to see them at Marchantaria? How did she recognize individuals? How did she follow them?

"At Marchantaria," she said, "I arrived at six A.M. and left at six P.M. each day, trying to identify and recognize the individuals. To recognize individuals is very important, very important," she said. "You cannot study them without."

So how, I asked, did she finally succeed?

"I tried for a long time, but I couldn't. It was too demanding. This will not work."

Oh.

Most of her earlier work had been population surveys. Most of her behavioral work had not yet been published. And those data, she explained in lilting English stung with Brazilian Portuguese, had been collected at a different study site, a flooded forest reserve a thousand miles away called Mamirauá. There she had freeze-branded fifty-six dolphins, and later outfitted a dozen others with radio transmitters, allowing her to track them. At Marchantaria, though, not even Vera could identify a single individual.

But we could at least learn from her what didn't work and why. As we watched the dolphins together, she could explain to us what the dolphins were doing, to give meaning to what we saw. She could show us the techniques that perhaps did not work for her at Marchantaria but might work for us at another site. Or so I thought until she had phoned us at our hotel, as Dianne and I were dressing for the opera, to report that she had contracted viral conjunctivitis and couldn't come.

An epidemic of the contagious eye disease was sweeping INPA, she reported. In fact, when we'd visited Vera at work, several people were out sick. Now people couldn't get to work because they couldn't see to drive; and even if they could get there, it was impossible to work, because with conjunctivitis, you are constantly rubbing your eyes.

My eyes began to itch.

Vera would send her field assistant, Nildon Athaide, to be our guide. "He knows everything," she assured us over the phone. Unfortunately, he speaks no English, I pointed out with dismay; and it was now too late to find a translator. "There will be no problem," Vera said. "But you must be sure to do one thing: You must be sure to buy a frozen chicken. Nildon very much likes chicken. *Don't forget the frozen chicken!*"

So that is how I came to be babbling incomprehensibly to a flustered clerk at the market that morning, desperate to purchase a food item that I, as a vegetarian, would not eat, and that Dianne, who eats anything, would cheerfully go without. It seemed to me a chicken was the last thing in the world to bring on a trip where the house would be literally surrounded by edible fish. But that was precisely why, I supposed, Nildon wanted it: he probably ate fish all the time.

In fact, Nildon was so eager for his chicken that when I did not return to the truck with it, he came inside the *supermercado* to see what had detained me. *"Sua frango!"* ("Your chicken!") I exclaimed in distress. He proceeded to a freezer and carefully examined every chicken in turn before making his selection. Next the chicken had to be weighed. But the supermarket had only an electronic scale, and the power was out. We implored the clerk to estimate the bird's weight, but she was adamant: If the chicken couldn't be weighed, it couldn't be priced. If it couldn't be priced, it couldn't be sold. And besides, the cash register was electronic, too.

Leaving the priceless chicken behind, we trolled the streets in our INPA truck, searching for a *supermercado* with signs of electricity. At last we found a market with the lights on, secured the frozen fowl, and within five minutes arrived at Pôrto da Ceasa and the open sixteen-foot aluminum motorboat that would take us to the Meeting of the Waters and to the dolphins.

. . .

The Amazon was full of trash. As we motored away from the dock, we viewed the muddy water around us with growing despair. We saw spray cans, cola bottles, oil cans every fifty yards or so. Plastic bags floated like jellyfish; stray barrels bobbed like bloated carcasses. We looked back toward Manaus and its industrial district where towering buildings proclaimed Samsung, Sony, Honda, Shell. An iron factory spewed black from one smokestack, while another glowed orange. Speedboats, reared up like anxious racehorses, whizzed by, so we had to pause to weather their wakes.

Within five minutes, we came to the Meeting of the Waters. Tourists peered down from a party boat at the two wide, wet brushstrokes of coffee and cream. From the high deck of their big ship, the tourists would have had a dramatic view, farther away from the garbage. The separate waters looked like living creatures traveling together, two monstrous eels or fringed water snakes, flowing side by side for nearly four miles. Atavistically, I wanted to touch the waters, as a child reaches out to

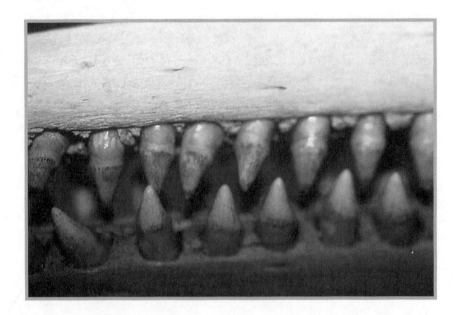

touch an animal; but I recoiled at the floating trash. It seemed impossible that dolphins could be living here.

Just then, two triangular fins split the waters. They sliced precisely between the two halves of the river, at the intersection of the two colors, as if being born.

"Tucuxis," Nildon announced over the roar of the fifteen-horsepower motor. In Brazil, these small gray dolphins are still called by the name the Mayanas Indians gave them in the Tupi language. We recognized them as the species scientists call *Sotalia fluviatilis*—the other Amazon dolphin that shares these waters with the boto. But unlike botos, tucuxis look and act the way we expect of dolphins: with their neat, compact bodies, short, well-defined snouts, and triangular dorsal fins, they launch out of the water, leaping and spinning, leaving arcs of spray as they spurt along the water's surface. Perhaps fifty yards from our boat, first one leapt, then the other, revealing soft, pinkish bellies; then the two leapt together, almost touching. Dianne and I grabbed each other's hands. "First the symphony," she yelled at me over the motor, "then the opera—and now the ballet."

Everyone likes the tucuxis, Vera had told us back at INPA. River people tend to be suspicious of the big botos, who approach boats so close and so suddenly. But the tucuxis are not as bold. They perform their joyous leaps at a distance, and they are small and pretty. Only four to five feet long, tucuxis look like miniature marine dolphins, elegant and streamlined, their bottle-snouts split with cheerful smiles.

Within the whale order, which includes the dolphins, *Sotalia* is classed in Delphinidae, the same family as the marine dolphins who swim in the seas and perform in oceanaria. In fact, until relatively recently, tucuxis almost certainly *were* exclusively marine dolphins, for even today they can be found in both fresh and salt water, ranging from southern Brazil to Honduras. Although they share the Amazon with botos, like

The strong jaws and conical teeth of the pink dolphin.

the black water of the Rio Negro and the white water of the Solimões, the pink dolphins and the gray tucuxis arose from separate origins.

The Delphinidae, comprising some twenty-six species, are a modern group. The most abundant and varied of the whales, they are compact and athletic, designed for speed-swimming in open waters. Although there are no fossil records of *Sotalia*, most scientists agree that these dolphins entered the Amazon from the Atlantic, probably no earlier than 5 million years ago.

But the botos are representatives of a very different whale lineage. Until recently, botos were classed with the other five species of river dolphins in the Platanistidae, the family to which the dolphins we had seen in Bangladesh and India belong; but now many scientists believe that boto and one related species, the La Plata dolphin of southeastern South America, should make up their own family, the Iniidae. Dianne and I had only seen botos in photographs and television documentaries, but even these images conveyed something eerie and ancient, a feeling you don't get from marine dolphins.

The boto's big body, which may stretch to eight feet long and weigh four hundred pounds, is quite different from most dolphins'. It lacks a prominent dorsal fin, possessing only a low ridge along the back. The flippers are huge, almost like wings. But it is the face that is most arresting: compared with the tucuxi's neat, smooth head, the boto's bulbous forehead seems misshapen, like a troll's or a dwarf's. The eyes are tiny. The face ends in a tube-shaped beak, which often curves to one side as if it has gotten bent. American scientists David and Melba Caldwell, who studied captive botos in Florida for many years, described them as "beady-eyed, humpbacked, long-snouted, loose-skinned holdovers from the past." But there is a strange beauty to the boto, a beauty that takes longer to see: it is like the beauty of the very old and the beauty of the fetus. Theirs is the beauty of becoming, of a creature poised on the brink of becoming something else.

Their lineage probably arose in western seas. Some scientists believe that the ancestors of today's botos may have entered the Amazon from the Pacific as long as 15 million years ago, when the Amazon still flowed westward into that ocean, before the Andes Mountains were born. The Amazon and the boto grew up together, at the beginning of the world. Back then, the South American plate was still drifting westward on the earth's mantle, on a collision course with the Nazca plate (which now underlies much of the eastern Pacific). The two plates' collision pushed up the Andes, and the mountains' rise cut off the great river from its only outlet to an ocean, creating the splendid isolation that allowed the Amazon to evolve its astounding diversity. The Mississippi, for instance, hosts only 250 species of fish; the Congo, fewer than 1,000; but the Amazon harbors more than 2,500 species, more than any other river in the world, and most of these species are found nowhere else on earth. For no other place on earth was like this: Until its waters finally managed to excavate the present, eastern course to the Atlantic, the Amazon, for 5 million years, was no longer a river at all, but one giant, flooded forest, the largest lake that has ever existed on the face of the planet. And each year, with the rains, that ancient, brimming lake re-creates itself. Now, just past the Meeting of the Waters, we would see the lake-world remade.

In less than thirty minutes of travel, all traces of Manaus and its garbage were left behind. Even the feeling of flow, that feeling of a river journey, was now behind us as we turned from the main channel toward a place called Catalão, where we would spend the next few nights. Catalão, it seemed to us, was not a town or even a village; just a stretch of reddish water punctuated by fruit-laden trees standing halfway to their crowns in the water.

Swollen by rains that had fallen almost daily since December, rivers throughout the Amazon may rise six to ten feet a month until the rains stop in May. Now, in April, in some places the water penetrated up to thirty miles inland from either bank, flooding thousands of miles of forest and village. Beneath our boat, we knew, saplings still grew green leaves under the water; and that night, while we slept, dolphins would swim beneath the floor of our floating house.

Besides our three-room house, painted gaily with red and green, and INPA's deserted double-decker laboratory boat, the *Harold Sioli*, we could see only one other floating house, a few hundred yards upriver, with a television antenna perched atop its aluminum roof. The television must have run off a generator, for there were no power lines. Like ours, this house floated on log pontoons and was tethered to a tree. The house belonged to a family of woodcutters, Nildon explained. We could see little gardens of potted plants growing on the porch of the house, where the people's chickens pecked at bugs.

Once Nildon cut the motor, the silence and the noon heat were stupefying. On the kitchen ceiling, an inch-and-a-half-long wasp was building a tubular nest. Tiny ants streamed over the kitchen table, undeterred by the water-filled cutoff Pepsi-Cola bottles in which the table legs stood as if in shoes. Outside, big green dragonflies perched on the purple blooms of water hyacinth beside the dead body of an eighteen-inch baby caiman. Other than insects, little moved. The three resident dogs and their four puppies, having greeted us, now lay on the floor as if in a swoon. Not even time could move in heat like this. So, as Nildon tucked his precious chicken into the little gas-powered refrigerator, we hung up our hammocks and waited for the noontime heat to pass. The water around us did not seem to flow. Instead, it sprawled all around, timeless, dark, shining and warm, full of promise and mystery.

Though we were eager to look for botos, the reasons to wait were sensible. Vera had found that, like most animals, botos are more active at dawn and dusk. Baking in the open aluminum boat at noon would be futile. Our limited petrol would better be used in the late afternoon.

We went out at two forty-five. By then, the heat had softened; it had grown no cooler, but more moist. Fat white clouds had gathered overhead, piled so high that the ones on top appeared to be squashing the ones beneath. We wondered when it would rain.

Shortly after three, the first tucuxis erupted from the water. First

one, then another, then two together, about seventy-five yards to our
starboard. I had read that they like to travel in groups of two or more;
the pink dolphins were said to be more often sighted singly. Nildon cut
the motor so we could watch their water ballet: Again and again, their
sickle-shaped fins sliced through the water, side by side. Another pair
came to join them. Perhaps they were hunting cooperatively for fish,
one group herding the prey toward another. Or perhaps they were simply enjoying the rush of water against their skins. A single tucuxi leapt
into the air, spinning, then crashed into the water with a splash. Large
whales do this to dislodge parasites, I had read; but the tucuxis appeared to be doing it for pure pleasure.

But where were the pink dolphins?

"Inia *gosta lago,*" said Nildon. ("The *Inia* like lakes.") This was why
Vera, who had initially begun counting *Inia* at the Meeting of the Waters, had moved her study to the floodplain lake at Marchantaria,
which we would visit tomorrow. Although both species are often found
together, the *Sotalia* prefer deeper waters, Vera had explained, while
the *Inia* prefer the shallower lakes and flooded forests. That is why their
backs bear only a low, subtle ridge: "You would be constantly bumping
your dorsal fin on sunken branches," she had explained.

But this feature also makes the botos far more difficult to spot. They
seldom leap like the tucuxis. If they were like the dolphins we had seen
in Sundarbans, they would appear to us first as only a motion—distant,
subtle, sudden. So we cast our vision wide, like a net, and hoped to
catch, in it, the signal of movement.

Then, at three-fifteen I saw it: a low gray fin, a bright pink back, the
color of sunset against storm clouds. It belonged in the sky, but was in
the water. That pink creature coming out of a river was impossible, my
brain said, and although the sighting lasted over a second, it was several long moments before I understood what I had seen. By then, the
boto was long gone. I did not even have time to call out, "Look!"

But Dianne had seen it, too, and of course, so had Nildon, who was
smiling broadly in the back of the boat, by the stilled motor.

We continued to watch the spot where it had appeared, as if the location had produced the dolphin and might offer up another. And indeed, two tucuxis rose out of the spot in tandem just minutes later; they rolled and jumped at the same spot again and again. Perhaps here was some sort of upwelling, I imagined, which attracted fish. Perhaps the tucuxis had been watching the boto, and learned from him where the fish were; perhaps the two of them chased the boto away. I longed to ask Nildon, but did not have the language skill; I longed to ask Vera;

The author watching for dolphins.

but most of all I longed for another glimpse of the boto—his tail, his flippers, his face.

We counted each sighting: in those afternoon hours, we saw tucuxis fifty-four times. There were perhaps ten of them, Nildon said. *"Onde fica boto?"* I asked plaintively. ("Where is the boto?") *"Nada,"* Nildon answered.

At four twenty-five, thunder rumbled in the distance. We had forgotten to watch the sky, and now we looked up into an aerial land-

Dianne Taylor-Snow with camera
and cigar. (Photograph by Sy Montgomery)

scape of white light over black clouds over cerulean air. Black and white swallows swirled like breeze, and we could hear the voices of other birds beginning their evening songs. But the sky itself was still, hushed, waiting.

We watched the sky in the water, as the sunset spilled itself over the wide river. Three tucuxis leapt out of the sunset. And then another low, pink form, a ripple made flesh, rose one hundred yards or so off the bow. We thought we could see the hump of the forehead, and before it dove, we heard the whoosh of breath. Another boto rose nearby—possibly the same dolphin, possibly a companion. And behind us, another breath, long and slow. We turned to see a wake. Next, to starboard, very close, mistlike spray. "Sometimes you see the breath, but not the boto," Vera had told us. It had sounded impossible, but now we saw it was true. Another gasp, from farther away. There were at least two of them. They had come to us.

The thunder came closer, too. A bolt of lightning cracked in the western sky, a stepladder of energy from sky to water, water to air. The lightning reflected the sunset. We had never seen anything like this: lightning a searing, impossible pink.

By six, the water had turned a molten silver, and the world looked like a photographic negative. Pink lightning throbbed on all sides. Later, I would recall with amazement that we were on a river in an open boat made of aluminum in a lightning storm and were completely unafraid. For we could have imagined no omen more obvious, more miraculous than this. The skies flashed pink, back and forth, like a message, a summons, an agreement, while invisible dolphins surrounded us with the moist promise of their breath.

UNFATHOMABLE FRAGMENTS

Mornings and evenings, we found, were the best times for seeing botos, just as Vera had said. This is true of many animals, who take advantage of the cool of the day, and the transition from dark to light, to forage, to play, or to hunt while their prey is active. But to us, it seemed as if dawn and sunset called them. Like a beautiful woman is drawn to her own reflection, the pink forms rose from the water as if to greet the clouds that glowed pink with the rising and setting of the sun.

They surfaced all around us. We would hear a breath, a splash, and we would spin around to see patches of rose and dusk skimming the water's surface. Sometimes it seemed they would pop up through the water like corks; other times it was as if they slowly gathered themselves from the water itself. And then, after just a second at the surface, they would sink from sight. We would stare at the wake, transfixed, as if it were an opening through which we might see the vanished animal, as if somehow the ripples, like one of those Magic Eye images, would suddenly fly into focus and make sense—but of course this never happened. And because we were staring at the wake, we would nearly miss the next sighting, which might be one hundred yards away, or right at the side of our boat—always, it seemed, in the direction opposite from where we had last seen one. The only thing we could tell for sure was the difference between a boto and a tucuxi: if we could tell what it was, it wasn't a boto.

None of these sightings were anything like watching marine dolphins in aquaria. In a clear-water tank, you can always see the entire animal; and besides, the trained Atlantic bottlenose not only leaps obligingly out of the water for its audience, but often suspends itself halfway, its head and upper torso visible, like a person treading water in a pool. Neither did our glimpses match the images of botos we had seen in films, photos, or illustrations. These, too, either showed the entire animal or at least a large portion of the body. Many of the photographs were of animals that had been hauled out of the water in nets. One dolphin, which we'd seen in a book in Vera's office, was actually lying on a mattress. Once we got to the floating house, we noted its fabric matched exactly those of the mattresses we were sleeping on. The documentary clips, I later discovered, were all filmed in semicaptive situations, when the dolphins had been cordoned off in clear shallows.

But to us, the botos came only in fragments: Once, a pink tail waved at us as clearly as a person hailing a cab. Another time, about a hundred yards away, a boto raised his head out of the water, spy-hopping like a huge California gray whale. But mostly, all we saw was the top of a head, or the curve of a dorsal fin—not an entire creature, but a flash of color and shape. We had to learn how to see them, to develop what is called the search image—to process that these pieces really were dolphins.

We looked desperately for markings that would allow us to identify individuals. This is a standard first step in studying animals, and often discovering the key characteristic to look for represents the first breakthrough. In studying mountain gorillas, Dian Fossey had found that each animal had an individual nose print. Studying African elephants, Cynthia Moss looked for distinctive notches in the ears. Tigers have individual patterns of stripes—and, some Indian researchers believe, individual footprints. Vera had tried to look for scars or notches on the dorsal fins of her dolphins. When a female is in estrus, males will fight viciously with one another, Vera explained, so there is ample opportunity to leave scars. So on our second evening, when we saw one dolphin with a particularly flat dorsal fin, and another with a V notched into his, we were overjoyed. But botos do not show their dorsals on every dive. In fact, they never seemed to reveal to us the same angle. We never saw those two again—or if we did, we couldn't recognize them.

Neither could we recognize animals by size. Sometimes, when an animal dove in an arching roll, we could tell when we were watching a particularly big or small dolphin. But botos don't always dive this way. They can float up and sink down, the way helicopters fly, and in these cases only reveal the melon of the top of the head or the dorsal ridge. Once, we knew we were looking at a youngster: the head was so small. He would have been about a year old, as most births here occur from May through July, when the waters are highest or falling. But generally, we had no idea of size or age, because such a tiny fragment of the animal was visible at a time, and for only a short moment.

The only clear feature of each sighting was color. Sometimes we were treated to a particularly vivid flamingo pink; other times, all we saw was a dark gray; many of the dolphins appeared to be mottled. But since we often had to look directly into the sun, and we almost always saw them at times the skies were changing, we knew even this could be an artifact of light. In fact, the dolphins actually do change color, though no one knows why: they may glow pinker with exertion, or with age, or their color might change with the temperature and turbidity of the water. "I know what our problem is," Dianne said to me as the creatures seemed to morph before our eyes. "These aren't dolphins. They're chameleons."

No wonder so little is known about them. "Although *Inia* was first collected by Alexandre Rodrigues Ferreira about 1790, little knowledge has been obtained since then about its biology," Vera and her husband wrote in a paper detailing the first seven years of their work. Theirs was the first, and still considered the only, major ecological study to be published on the species. From pooling data with biologists studying botos in Colombia, Peru, Bolivia, and Ecuador, and from themselves patrolling thousands of miles of river at a leisurely seven miles per hour, the couple determined the boto's range is enormous. The species is found in most of the major tributaries of both the Amazon and Orinoco, with the exception of the lower reaches of the Xingú River at the mouth of the Amazon and the southern half of the Tapajós. And when the wet season floods the land, the dolphins enter every riverine habitat in the Amazon basin, from grasslands to treetops.

The couple also collected carcasses of dead botos. They measured them. They dissected the bodies. They measured the organs. They counted the parasites. Vera was particularly eager to look in their stomachs to see what they ate; a 1970 report claiming that botos compete with fishermen had sparked her initial interest in the species. Less than half the fish they eat are commercially important, Vera found, and these are taken in relatively small quantities. For botos can eat almost anything: with big, conical teeth and strong jaws, a boto can break a turtle shell. In several stomachs, she has found rock-hard armored catfish with erectile spines. Botos will take some fifty species of fish, and in a single meal might consume more different kinds of fish than other species of dolphin might eat in a lifetime.

But, as important as these findings are, Vera's papers still detail a litany of unknowns: "Virtually nothing is known about the physiology of *Inia* . . . its overall abundance is unknown . . . where these animals go during the dry season is unknown . . . there has been little research on *Inia* in most of the countries where it is found . . . there are insufficient data to be analyzed for trends for any population. . . ." No one knows when the breeding season is, if there is any. No one knows the structure of social groups, if they exist.

At the Meeting of the Waters, Vera found she could not even begin to chronicle the dolphins' lives. "They are not easy animals to study in the main river," she explained to us later. "It's too wide." The animals seemed to flicker at the edge of our consciousness—a movement at the corner of the field of vision, a spume of breath that lingered after the animal dove, fragments we could not hope to unite into meaning.

But occasionally, a glimpse called to us, like a snatch of vaguely re-membered melody. One day at 5 P.M., after Nildon had stilled our boat by wedging it into the crown of a drowned careru tree, three of them came. Two swam close together; a third seemed to have arrived alone. In Vera's research, most of the botos sighted were solitary, and groups greater than two were seldom encountered. Twelve to 26 percent of the observations Vera analyzed were pairs like this one: probably, she con-cluded, mother and calf. The bond between mother and offspring is very close; Vera has found they stay together for at least two and a half years. We had seen photos of mothers and young swimming together, often one on top of the other, their long snouts touching, one flipper trailing along the other's flesh, the way a child holds a parent's hand simply for the comfort of touch.

Four times the two botos surfaced within fifty yards of our boat, inches from one another, and I imagined—though I could not see—that they swam touching. For they are sensuous animals. An observer at Germany's Duisburg Zoo, which has housed two botos since 1975, recorded the dolphins blowing curtains of bubbles through which one or the other would then swim; in one case, the elder dolphin, Vater, grabbed a scrub brush, left in the tank as a toy. Pulling it quickly down-ward through the water in his mouth, he created, along the whole length of the handle, a rising curtain of prickling and glittering air bub-bles. The younger dolphin, Baby, then rose in the water and rolled in the bubbles, "obviously with great pleasure. . . . The purpose of the air bubbles," wrote Wolfgang Gewalt, "seems to be the benefit of a special kind of caressing massage, similar to whirlpools."

In captivity, botos are inventive lovers. After foreplay in which the male nibbles gently at the female's flippers and flukes, they make love in at least three different positions: head-to-head, head-to-tail, and at right angles; they may indulge in this forty-seven times in three and a half hours. Both sexes masturbate, the male rubbing his penis against objects, the female inserting objects into her vagina or pressing against it. Males sometimes use the sensitive penis as an exploratory probe, even inserting it into another dolphin's blowhole. The botos' pursuit of sensual pleasures, Gewalt wrote, seemed "more highly developed than any similar behavior known up to now from other cetaceans, either free living or kept in the Zoo."

And then we saw the young dolphin slowly roll in the water. Languidly, the baby turned, revealing a chubby pink belly and paddle-shaped flippers spread akimbo. For a long moment, with his mother beside him, the yearling floated at the surface on his back, as if he were savoring the feel of sun warming his skin, and then, equally, the way the river reclaimed his body in its cooling welcome.

How I wanted to join them! Vera had spoken with us of this same longing: "Sometimes it is my desire to swim with them," she had told us in her office at INPA, "but there are piranhas. The water looks nice, but I always think of the piranhas." She has friends who have been badly bitten. She has found botos with crescent-shaped hunks taken out of the skin, the work of razor-sharp teeth. "And when we catch the boto to freeze-brand, we make a circle with our net," she had told us. "We pulled the net out—this new net, a beautiful brand-new net—and it came out like a rag! The piranhas destroyed it!"

Anything might be in these waters, she had warned. "You may feel something touching you, but you do not know what it is. It can be a piece of wood, it can be a grass, it can be a snake, it can be a piranha, it can be a stingray—it can be a shark. We have caught in the Solimões River four huge sharks—two meters and a half. And it's a marine shark! It comes in the Amazon and can be found all the way to Iquitos in Peru. They're almost the same color as the water. When they come here, they get the mud on the skin. So I will not swim with dolphins!"

Never before had I felt so separate from water. There was us; there was the surface; and there was the impenetrable world below. *Anything might be in these waters.* I remembered the legends about the boto's enchanted underwater city, which people deeply fear. But their fear is tinged with desire. I had read that shamans will sometimes seek out dolphins, contacting them in dreams or in ecstasies induced by hallucinogenic drinks or snuff, and ask them to show them the secrets of the Encante. The people know the river is dangerous. But beneath the river, they say, is a place of unimaginable riches, the treasures of lost souls, pleasures to quench every desire. The dolphins there preside over a world where there is no longing, only music and singing and dancing. Sometimes, it is said, you can hear snatches of their songs.

Sometimes we heard strange sounds underneath us. The bottom of our boat was a stethoscope to an unknown beneath. Early one morning, we heard the high-pitched whine of an electric current. What could it be?

"*Pescada,*" answered Nildon. ("Fish!")

Then we saw one dolphin take off like a shot beside our boat. "*Muito rápido!*" exclaimed Nildon. "*El boto fazer compras.*" ("The boto is shopping.") He was chasing a fish.

After we had watched them for an hour, the dolphins disappeared at six thirty-four that morning. We waited and waited. Kingfishers zoomed like arrows across the brown water. An electric green parrot screamed from the top of a tree. We listened to fruits falling into the water. But we saw no more dolphins, heard no more breath.

By eight forty-five, the heat was gathering like a cloud of gnats. The dolphins did not return. So we headed to Marchantaria, the floodplain lake named after a grassy island there, six miles upriver along the Rio Solimões.

Vera had picked five checkpoints from which she tried to identify individuals as they entered and left the lake. This time of year, Vera had told us, the botos come to the lake following fish, who go there to eat the fruits dropping from the trees in the flooded forest.

Our days and nights on the river were punctuated by the splash of the ripe fruits the trees released into the river. In the rainy season, there is fruit for everyone: the monkeys and parrots, the fruit bats and bamboo rats, the dainty brocket deer and the strange, snouted tapir, a four-toed forest dweller who looks like a cross between an elephant, a rhino, and a pig. But the trees do not make their fruits for birds or mammals to eat and disperse, as most fruit trees do. To carry their seeds to the locations where they will grow and feed these many mouths, the trees here make their fruits for fish.

Nowhere else on earth do fish depend on seeds and fruit for a major part of the year. But in the Amazon, the trees release their seeds to voyage on the water, and more than two hundred species of fruit-eating fish migrate into the flooded forest to gorge and to spawn. The fruit embarks on a journey of rebirth, and its splash is the first note in an unheard symphony of accommodations between water and trees, trees and fish, fish and mammals.

The trees here have evolved ways to make their fruit attractive to fish and easy to find. Most fruits float. Many of the fruit trees, such as laurels and cinnamons, produce fragrant organic latexes, oils, resins, and acids. The fish wait beneath these trees, drawn by the scent. Some,

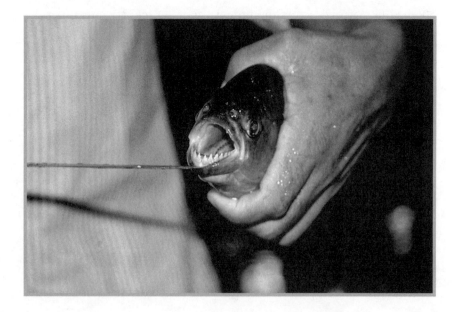

like the tambaqui, have developed nasal flaps on the upper part of the snout to help them smell these fruits. To crush the seeds of the seringuera tree, tambaqui have molars like those of horses, and huge, powerful jaws. Others, like the ancient, black-armored catfishes, swallow the stonelike seeds of palm fruits whole and digest the fleshy covering. The seeds pass through the gut and are defecated whole, where they can germinate in a new location. There, once the waters recede, they do not have to compete with the parent tree, nor attempt to fight its shade.

"It is tempting to hypothesize that fish and trees have evolved in a mutualistic relation," Michael Goulding, one of the leading experts on Amazon fish, has written. The sustaining relationship between fish and trees here has certainly been noted by the people who have lived for centuries along the rivers that flood these forests. One of the central myths of Amazonian people is the story of the World Tree, which in many river cultures is said to be full of the water that gave birth to the Amazon and its fishes. As a Shipibo myth tells it, the world began this way: When the sun first emerged, it hit the branches of the World Tree, which released its luscious fruits into the lake like rain. Fish rose to the surface to eat the fruit. From the bites the fish took emerged all the species of birds in the world—connecting the water with the sky.

And at the edges of the day—at the times when dolphins come, at the times the clouds gather—the world re-creates itself from water anew. The scientists now tell us that life itself lifts up from the water, like a spirit, to seed the clouds. Oxford and East Anglia University scientists have proposed that a gas emitted by many algae—dimethyl sulfide—evolved to allow small tropical microbes to rise from the water and take to the air. And this same gas, they note, is well known as the primary natural source of airborne particles around which water droplets form to create clouds. The organisms do this, suggest the scien-

Piranhas' sharp teeth can reduce a fishing net to rags.

tists, for the same reason trees make fruit: to speed them on their journeys toward new life. In the heart-quickening thunderstorms that soaked our nights in Brazil, these tiny lives were riding the rain back to earth.

The legendary Greek philosopher Thales of Miletus, whom Aristotle considered two centuries later to be the founder of the physical sciences, believed that all matter consists ultimately of water; that is easy to believe here. Despite Vera's warning, I put my hand into the soupy lake: it feels thick and alive, not just fluid, but tissue, like skin and blood. This is not merely a medium for life, I thought: the water *is* life, itself alive.

Travelers have long written of being overwhelmed by the immensity of the Amazon. "A vast inland sea," as explorer Henry Walter Bates called it, the river stretches 4,000 miles, discharging fourteen times the daily flow of the Mississippi. Its jungle spreads for 2.5 million square miles—"a forest the size of the face of the full moon," anthropologist Wade Davis observes. But to me, the immensity of the Amazon was here at my fingertips, as I put my hand into the living water. The immensity of the Amazon is in its unfathomable wholeness, a wholeness forever giving birth to a universe unimaginably strange and perfect and unseen. And this, I knew, was the source of my longing to follow the dolphins: though I was drawn by the fragmentary glimpses they had shown me at the surface, I was compelled by their siren strangeness, their immersion in the wholeness of the unknown world below.

After forty-five minutes' travel, we arrived at a flooded meadow. The tassel-like seedheads of the grasses smelled sweet as orchids. Egrets flashed bright against the sun. There is another INPA floating research station here, and we stopped by to speak with some visiting German scientists to see if they had seen botos. They had seen two that morning, and more would be coming soon: as the rains continued, the water was expected to rise another meter in the next week.

We journeyed farther, the motor at a low purr as we scanned the water. Each small wave seemed to gather itself into a promise, seemed to darken and quicken, as we prayed for a pink fin to arise, like the limb of a baby breaching from a dark womb. But then the wave would dissolve into itself, and hope vanished as completely as a fruit swallowed by a fish.

The sun was so hot it stung our skin. Sweat poured into our eyes as we squinted into the glare. Each night, our sleep had been fragmented by lightning storms and downpours. As Dianne swore in various languages, we had had to get up and secure the shutters in the middle of the night. Once awake, we then realized ants were crawling on us, a fact that had mercifully been concealed while we were asleep. So we were exhausted as well as hot.

My Portuguese was so poor I couldn't really converse with Nildon except to ask, *"Onde fica boto?"* ("Where are the dolphins?"), which I realized must have sounded like a complaint. Nildon, who was eager to please us but who could not produce dolphins on demand, would shake his head and reply with disgust: *"Não há boto aqui hoje."* ("There are no botos here today.") He piloted the boat over to the island, where we saw two other boats and the hunched-over forms of five men working with nets.

They were a team of electric fish experts from INPA, two of whom, to my delight, spoke English. A hefty young man in a wide-brimmed Panama hat introduced himself as José Gomes; wiry, thin Paulo Petry was wearing a camouflage hat and horn-rim glasses. They were bent over a big black net, hauling out wet masses of water hyacinth as glittering dragonflies swarmed around them. Periodically, José inserted into the weeds a long bamboo pole with a voltmeter attached. It crackled with the current of electric fish.

"Each species has its own signal," explained José. The signals sent from the electric organs, which are similar to muscle tissue, are picked up by the electric receptors of other individuals. An individual fish can alter the shape of the electric field, the waveform of the discharge, the discharge frequency, and the timing patterns between signals to send different types of information.

Had we heard fish signaling one another beneath our boat that morning? What were they saying to one another? Experiments in captivity show electric fish can convey to others their species, their sex, and their age class. They give territorial warnings, like birds. In some species, the broadcast can carry thirty-two feet. "The behavior is incredible," said José. "They use these signals in social interactions, to detect objects, for dominance, to find mates—there are so many things we are trying to understand!" In the Encante beneath the waters' surface, in the language of electricity, these tiny fish sing songs of lust and fear and courage—operas played out on a vast underwater stage.

These fishes' current isn't dangerous to people. Unlike electric eels (who aren't actually eels at all, but are themselves a kind of fish), they emit only a fraction of a volt. But Paulo called out, *"Cuidado—cuidado!"* ("Careful!") A five-inch catfish was embedded in the tangle of vegetation, and it erects three toxin-tipped spines to turn itself into a poisoned trident spear when threatened.

The INPA team, José explained, was looking for a particular species of electric fish belonging to the family Rhamphichthridae. They are transparent, shaped like streamers, long thin triangles with no back or side fins. "We didn't find the one we are looking for," said Paulo. They knew from the voltmeter that this family was not in the net. But instead, there was a surprise: they had come up with an eel unknown to science—a foot-long creature appearing to belong to a different family, Symbranchidae, whose pronounced jaw gives it a triangular head like a viper.

"Last year there were only two new eel species reported in all of South America," Paulo said, shaking out the water weeds. "There were five new ones from this one island in the past two years!" Among them is a species that makes nests and burrows in the mud in the dry season. They wait out the dry season in a state of suspended animation, until the water floods in and brings them back to life.

The Amazon's bewildering diversity is unheard of anywhere else on earth. Michael Goulding, who had generously offered me advice on watching botos, has single-handedly discovered two hundred previously unidentified fish species in the two decades he has worked in the Amazon. An acquaintance of Dianne's, Dutch primatologist Marc van Roosmalen, had gone looking for a new species of marmoset he had heard about a year ago. In the process, he reportedly discovered four other previously unknown primates, a new species of dwarf porcupine, and spotted what he believes to be an undescribed tapir and an unknown subspecies of jaguar! All of them, incredibly, were discovered barely 190 miles from the city of Manaus.

How to even begin to fathom such richness and complexity? What was I even doing here, trying to follow an animal I could not see into a world I could never know?

Here, drenched in the ferocious light of the Brazilian noon, I understood why "fathom" is a water word, one we use when we are trying to understand something difficult and deep. In fact, a dictionary definition is "to determine the depth of." Another definition of "fathom," the verb, is "to sound." Which is, of course, one way the dolphins penetrate the meaning of their world.

Their eyes, though they appear small, are actually, in the skull, as large as those of marine dolphins, and they function quite well in air and in clear water. But in the brown murk of the Solimões, the botos rely on their sonar, throwing out a beam of sound to chart their world with its echoes. This is one reason the botos need such flexible necks: as they swim through the flooded forest, they turn the head from side to side, sounding their way through the maze of drowned branches with a series of pulsed clicks at frequencies up to 170,000 hertz, far higher than we can hear. Ordinary sounds would be too inaccurate on account of their longer wavelengths and would not produce a precise enough echo.

One researcher has compared the sonar system of river dolphins with an electron microscope: the resolution of an ordinary microscope ends where the object is smaller than the wavelength of light, whereas the electron microscope uses a beam with a shorter wavelength than that of light to detect minuscule items. The boto's sonar is thought to be the most highly developed of all. Alone among the whales, they can fathom the flooded forest.

So I could hope for no better guide to the Amazon than the boto. Dolphins have guided humans for millennia. Legends older than the Minoans say dolphins led the ancients to the center of the world—a sanctuary named Delphi, in their honor. The Greek sea god, Poseidon, turned to a dolphin to find his lost bride, Amphitrite—a miracle com-

memorated by the dolphin constellation in the sky. Dolphins carried the souls of the Etruscan dead to the Isle of the Blessed; dolphins symbolized the Christian soul's rebirth.

I looked to the botos for no less of a miracle. Why was I drawn to follow them? Of course, I wanted to see where they go. But in my heart, I wanted to follow them deeper yet: I wanted to follow them to the Encante—to the city beneath the waters. I wanted to follow them to Eden. I wanted to follow them back, down, deep, to the watery womb of the world, to the source of strangeness and beauty and desire. I wanted to touch the very soul of the Amazon.

The concept of under-standing implies probing something deep, something beneath, a search for a beginning or origin. The desire to fathom the world is perhaps our deepest human longing. The word "fathom" originates in the Old English *faethm,* meaning "outstretched arms"—a gesture, I realized, of divine supplication.

And if math is the language of God, as the physicists say, it is no wonder that a fathom is also a unit of measure: six feet. Measurement is one way science tries to penetrate mystery: We measure voltages. We sort stomach contents. We count botos.

But at the Meeting of the Waters, and at Marchantaria, not even Vera—much less Dianne and I—could reliably count botos. So the morning before we returned to Manaus, I measured the one thing I could count: their breaths. As what we believed to be two botos surfaced around our canoe, I counted their breaths. They breathed approximately every two minutes: 6:06, 6:09, 6:11, 6:13, 6:14, 6:15, 6:17 . . . and then they disappeared.

Right before we left, we saw their shadows pass beneath our boat. It was the closest we had come to seeing the animals whole.

Researchers from INPA discover an eel unknown to science.

Desire

"HERE IN THE AMAZON, THERE ARE TWO KINDS OF BOTO. SOME are enchanted," João Pena says. "There's a time in the still of the night when you can tell the difference between the boto and the encantado. A lot of people, people like you, even people from the Amazon, they say they don't believe it. They say, even if we are telling a story of something that really happened to us, that it's just a fisherman's lie. But there are people here that know. They know that it's real. It's really true."

He was twelve when he saw his first encantado. It was the night when the great turtles come out of the river to lay their eggs on the white sand under the full moon.

You must wait until they have finished laying before you take them, he explained, because you must ensure there will be turtles to eat the next year; so it was quite late by the time they began to collect their quarry.

"Just as we were picking up the turtles, we saw three men sitting in the sand, dressed in white. Why would they sit when there were turtles to be caught? Who were they? We went closer to see." But as they approached, the white-clad men sprang up and leapt into the Solimões. The boys watched for the men to emerge—but they never did. Instead, they saw three botos leap just where the men had disappeared. The botos came quite close to the shore, he remembers, chuffing spumes of spray through their blowholes, as if indignant.

"There are so many stories of botos, you will die writing!" João Pena says. In the floating house at Marchantaria lived a man whose teenage daughter was beset by the attentions of an amorous boto. The dolphin had a distinctive white spot on the tail, and he was always near the house. At night, he would come out of the water to visit her room and enter her dreams. He appeared in the form of a man wearing white shorts—with a spot on his leg. The boto never touched her, but she was very afraid. She felt his presence, his longing for her. Day by day, she grew more pale and anemic. One night, the boto appeared to her mother in a dream, and asked her to give the girl to him. She refused. But the boto kept coming back, night after night.

Finally, the family had to move from that place. Now the girl is twenty-five. She isn't bothered by the boto anymore, she told us. But to this day, whenever she visits the river, botos come near the shore, desiring her still.

João knows that the girl and her father are telling the truth, he tells us, for the same thing has happened to him. When he was seventeen, with his uncle, he had joined a fishing party that had gone to harpoon pirarucu. While they were on the beach gutting their catch, a boto came near the shore, looking at them. The uncle joked, "Pretty dolphin girl, are you looking for a handsome young man? See João Pena! Isn't he a good-looking fellow!"

That night, they slept in their canoes, anchored in the middle of the river. "The night was starry," João remembers. He was in a canoe right next to his uncle's. He felt the craft dip, as if it were taking on weight, and opened his eyes. Sitting in his canoe was a young woman. "Only, she wasn't dressed in white," he told us. "She was naked, with big breasts.

"I leaned toward her. She was so beautiful, with long blond hair, and her skin was very pale. I was enchanted. And then she went back into the water, and turned into a boto."

The splash woke everyone in the party. They looked around and saw their canoes were surrounded by pink dolphins! The men were frightened and, trying to get away, paddled their canoes into the forest. But the botos followed, splashing their tails as if they were angry.

Why were the dolphins following? "You see, the encantado, she had fallen in love with me," João explains regretfully. "The other dolphins were jealous. Botos are very linked with one another." The botos did not leave until the following morning, splashing all the while.

And what if he had followed the boto-woman into the water? What would she have done with him? Is there such a thing as the Encante?

"Oh yes—the Encante is very real," he says. "In the time of my grandparents, they talked of it. Across from Manacapuru, there was land with a lake and grass on top of the water that never died. In the middle of that lake you would hear cattle, dogs, bands playing, singing. All of us went walking toward the lake—my mother, my aunt, my grandmother. We all heard it. We thought it was a party. And in those days we used kerosene lamps, and we carried one over, to see who was having a party. It got louder and louder—there was so much noise! But we never saw anybody. They said, and I know it is true, that the sound was coming from beneath the water, where the dolphins live."

João Pena and the father of the girl who was visited by the dolphin both work for Vera at INPA. She depends upon and respects both men. But she does not believe their stories—and she has heard so many in her work. "Not far from Marchantaria, a woman asked if I was afraid to be enchanted," Vera told us one night back in Manaus. Rain roared over the tin roof of the open-air restaurant where we were sharing tambaqui com leite de coco, and lightning flashed on all sides. "She told me, 'Don't eat anything the boto offer you. If you eat, you'll never come back!' " Vera laughed her bubbling laugh. "I told her I would like so much to visit the Encante. And she was horrified!"

"But you said you don't believe in the Encante," I said.

"Yes," she answered, "but sometimes, I wish it were true."

IQUITOS

Actually, there is an underwater city. Two days after we left Manaus, we found ourselves canoeing through it.

The swelling river had claimed half of Belen: the first, and half the second, floor of the Bar Don Freddy; the bottom floors of the cement and steel dental clinic; the lower half of a giant billboard advertising Inca Cola; the dirt streets; the wooden bridges; the bottom floors of the stilt houses; the people's gardens—all were under water. Electric street-lamps poked up from the river, like the craned necks of the black water birds called anhingas. Farther away from shore, where the houses are made of salvaged tin and plywood and thatched with yarina palm, sticks thin as a child's wrist held tangles of electric wires powering naked lightbulbs inches above the water's surface.

No dolphins live here, but people do. During the months that the rains flood half their town with the river's wet embrace, some 40,000 people crowd into the topmost floors of homes and restaurants and little stores rising on stilts from the brown waters of the Amazon. Their chickens and pigs and dogs retreat up here, too, like residents of Noah's Ark, and people grow vegetables and flowers in potted gardens until the water retreats and they can descend to the ground once again.

Meanwhile, the people climb up and down wooden ladders to and from their canoes, and the youngest children splash in the river as if it were a suburban backyard, the girls' frilly dresses plastered wet to their skins. The older children, wearing blue and white uniforms, paddle to the secondary school. At recess, they play soccer in a classroom, because the soccer field is under twenty feet of river.

Iquitos, of which Belen is a part, receives 120 inches of rain a year, more than any other city in Peru. When authors Brian Kelly and Mark London visited it a decade ago, they found the place had "the feeling of a lifeboat, crowded with people and jammed against the jungle wall . . . one good rain, we felt, and the whole sodden city might slip away," they wrote. In the early 1980s, the city was shrinking yearly, as the Amazon, meandering westward, chewed chunks from the banks during the rainy season. (In the last decade, the river has begun instead to cut a deep channel to the east of the city proper.) But Belen, its drowning slum, grows each year, as poor people come from the countryside, seduced by the prospect of money and the promise of "civilization" in Peru's largest jungle city.

The underwater city of Belen is no Encante. Salesmen paddle door-to-door offering tomatoes, fish, cigarettes, cola, and juãnes, the ubiquitous rice dish cooked with chicken or pork and wrapped in ginger-scented leaves. They stop at the houses where the women are washing their laundry in waters over which their own latrines are perched—the same water in which their children are swimming, the same water they drink unboiled, despite the public health warnings billboarded everywhere: EL COLERA ES UNA ENFIRMEDAD QUE PODEMOS EVITAR! urges a UNICEF poster. (CHOLERA IS A DISEASE THAT WE CAN AVOID.) But of course, most of the people who live here, like roughly half the population of Peru, can't read, and some of those who can refuse to boil the water anyway. "But boiling it will kill all the life in the water!" one health worker recalled a client had protested.

Though Iquitos as a whole is said to contain 328,000 people, no one

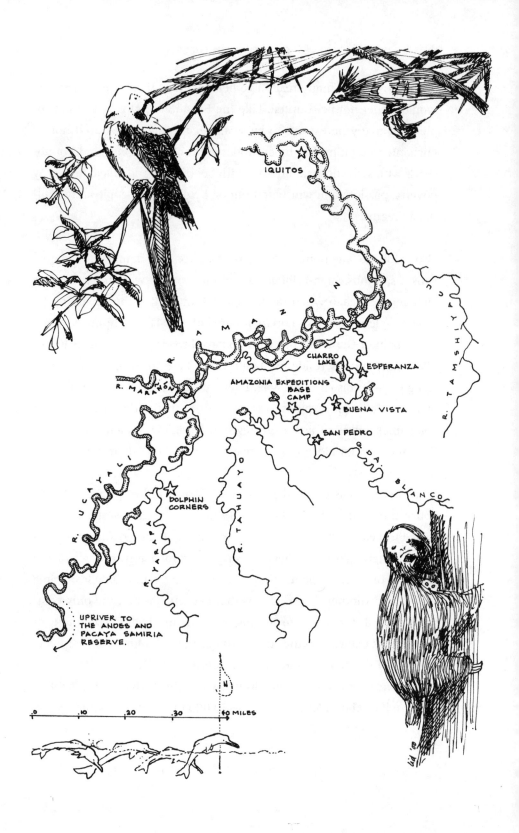

IQUITOS

R. TAMSHIYACU

AMAZON

R. MARANON

CHARRO LAKE

ESPERANZA

AMAZONIA EXPEDITIONS BASE CAMP

BUENA VISTA

SAN PEDRO

Q.DA BLANCO

R. UCAYALI

DOLPHIN CORNERS

R. YARAPA

R. TAHUAYO

UPRIVER TO THE ANDES AND PACAYA SAMIRIA RESERVE.

0   10   20   30   40 MILES

N

even knows how many really live here. Most of them are poor. Many would rather not be counted. Like the Encante, Iquitos swallows people—but many here want to be swallowed. They are seduced not by enchanted dolphins, but by an underworld of a different kind. Iquitos swirls with seductions, a muddy whirlpool of need and greed, driven by poverty, passion, and whispered rumor, by displaced people and unfulfilled dreams.

"No roads, no telephones, no radio—that's for me!" Jamie R., who now runs a spacious, high-ceilinged restaurant on the waterfront, decided eighteen years ago when he arrived in Iquitos. It is the capital of the huge jungle province of Loreto, an area of 14,241 square miles, larger than Spain or Germany, but with no paved roads penetrating its forest. Like many Iquitos residents, Jamie came here to escape from something else—in his case, his wife. Expat gringos tell tall tales in the bars about their jungle exploits, and workers with the dozen jungle tour companies and the thirty-eight registered nongovernmental conservation and foreign-aid groups spread rumors about their competitors' dark sins.

In the dry season, we'd heard, the place is full of gringos like us. Tourism declined a bit in the early 1990s when many western nations issued travelers' advisories warning of Shining Path terrorists and bandits (though there was never a single recorded terrorist act in the province of Loreto—mainly they stuck to the overland route from Lima to Cuzco). Another deterrent was that Faucett Airlines, the only company with direct flights to Iquitos from Miami, had a reputation for crashing its planes into the Andes. (Bankrupted by fines, the company folded shortly after our return flight from Peru.)

But since Peru was removed from the travelers' advisory list in 1993, tourism has rebounded. Today some 30,000 foreign tourists and 73,000

The underwater city of Belen.

Peruvians visit the region each year, for it is the gateway to Peru's Amazonian jungle. Its enchantment beckons adventurers of all sorts. Some come to kayak, some come to count birds; some come to meet Indians, some come to watch monkeys; some come to take home blowguns and monkey-teeth necklaces. Some come to immerse themselves in the keening calls of the jungle night. They come in search of forest primeval; they come in search of their souls.

Tour guides still tell of some of these seekers. There was the man from Wisconsin who insisted on wandering barefoot through the jungle, drinking river water, and sleeping without a mosquito net. (He emerged from his adventure with punctured feet, cramping diarrhea, and bug bites gone septic.) There was the self-styled "shaman" from California who came to sample the sacred drugs of the local Ayahuasqueras. (He spent most of his time throwing up, and in the end, begged to go home.) Some tourists keep coming back year after year. But Iquitos feels like a city where no one *lives;* and in fact a sizable portion of the population stays only a short time. Teachers, evangelists, aid work-

ers, soldiers, and sailors do their tour of duty and move on; traffickers in illegal drugs and contraband animals take what they can, get out, come back. "No one asks questions," Jamie says. "You can get anything you want here; you can be anything you want here."

Though it was founded as a Jesuit mission in the 1750s, since the rubber boom Iquitos has been the nefarious trade capital of the Peruvian jungle. The rubber barons who built their headquarters here were responsible for some of the worst atrocities committed against the rubber tappers: 30,000 Indians were murdered on the Putumayo River, just to the north, sacrificed to the rubber empire of the infamous Júlio César Arana. One of his overseers had as his motto, "Kill the fathers first, enjoy the virgins afterwards." Despite its mansions and tiled walls and riverside promenade, even at the height of its wealth, Iquitos harbored few pretensions to grace. Prostitutes filed their teeth razor-sharp like piranhas. On carnival days, section chiefs would toast with champagne the man who boasted of murdering the most Indians that year.

Exposing the cruelties of Arana in a series of newspaper articles for a London-based weekly, Walt Hardenburg, a twenty-seven-year-old railroad engineer from New York, hastened the end of the Amazon's rubber boom in 1913, inciting worldwide outrage at the atrocities. But today, from this bend in the river 2,300 miles due west of the Atlantic, Iquitos continues to bleed forth the stolen lifeblood of the jungle and its people: oil, timber, cocaine, animals.

Iquitos is still an isolated frontier town. Though it now has radio and telephones, there are still no road links with the world outside; you get here by air or by boat. We had flown from Manaus to the little border town of Tabatinga, where Brazil meets Colombia and Peru, and where we'd stayed in the tiny Hotel Christiana in a room swarming with mosquitoes. There were no hooks for mosquito nets, and its blue-painted cement walls were impenetrable to the thumbtacks Dianne had brought for such an emergency. We determined the mosquitoes were attracted to the lidless, seatless toilet in a doorless, low-lying corner of the room, which was surrounded by a small pond of leaking water. "You would insist that we come during the wet season," Dianne said. "It's high water even in our *room!*"

We'd set out at four-thirty that morning on the twenty-seat boat, the *Loreto Rápido*—so named because the boat's 230-horsepower diesel engine permits travel at the impressive pace of forty miles per hour, arriving in Iquitos ten hours later. We had come here to meet Moises Chavez, who works for a friend of mine, Paul Beaver. With business partner Suzy Faggard, Paul runs the only lodge in the area, on the edge of the 800,000-acre Tamshiyacu-Tahuayo Community Reserve on the Tahuayo River, a day's travel from Iquitos. The reserve and its surrounding area were astonishingly rich in wildlife, with more species of primates (14) and rodents (26) than any other area of South America—and there were plenty of dolphins, Paul had promised. He generously offered his lodge as our base, and arranged for Moises to act as our guide.

Moises met us at the garden of our cozy hotel, La Pascana, in Iquitos. A small, strong man of thirty-eight, with wide, high cheekbones and a voice soft as wet leaf litter, Moises' dark, dreamy eyes always seem to be looking past you, as if scanning the forest for a particular tree in fruit or a troop of monkeys or the cryptic form of a sleeping sloth. His father had been a teacher, whom the government sent to isolated tribes. With his seven brothers and four sisters, Moises grew up learning about the jungle from the Indians on the Napo River: the Yaguas, who decorate their faces daily with elaborate red designs and dress in grass skirts, and the Secoyas, who pierce their bodies and hang their fishhooks conveniently from holes in the nose.

"In Iquitos, many years ago, before the Belen market was built, Indians lived there, on the bank of the river," Moises told us. "What happened was, many young girls and young mens celebrated on Saturdays. But it got out of control. Everyone's gonna drink beer. The dolphin watched this for one year. And when this one girl comes to the river, that dolphin, he's gonna watch her. But she doesn't know. Some people say the dolphin's just a dolphin, but the Indian people, they know a different story."

Moises knows the stories well. He has heard dozens of them, over and over—from his father, from his grandmother, from his Yagua and Secoya friends. The beginnings of the stories are always long and convoluted, like the meandering canals of the Amazon, and his sentences, like the waterways, crisscross one another, weaving past and future, germane and incidental. He recites with almost poetic cadence. He recites them trancelike, with a faraway look in his dark eyes, as if seeing his way back in time.

"One day—a day like today—they gonna celebrate a really big party," Moises told us. "One day, an orchestra from Iquitos was going to play. . . ."

That night, when everyone was drinking and dancing, a handsome man with white skin and blue eyes showed up. The beautiful girl asked the young man to dance. "I've never seen you before," she said. "Where is your home?" He said, "I live here—I saw you one time when you swam by the beach near Iquitos." But the girl, said Moises, had never seen him.

She quickly fell in love with the stranger. And that very night, he asked her to marry him. "I will give you a present," he said, "a gold watch. My father is a rich man, and I have many gold items at home. I will invite you to my house someday." And that night, along with the gold watch, he gave her a fine diamond and a gold ring. She was very happy. But the young man warned her not to show her new gifts to her mother or brothers or father. "Hide them," he said. "They will be good luck for you."

"Till two-thirty in the morning they danced," Moises told us. "They planned to meet the next week, at the next dance. They said good-bye and he disappeared."

But the greedy girl couldn't resist boasting to her parents, her brothers, and all her friends about her rich new boyfriend, and all of them showed up at the dance looking to meet him. "Then," said Moises, "about three o'clock in the morning, they tried to restrain him. But he disappeared. One of the men heard a big splash in the water; but no one thought anything of it."

The next week, the stranger returned to the party, seeking the girl. This time, he brought her beautiful clothes with gold buttons, beautiful shoes, a golden necklace. They would be married soon, he said. Again,

Moises Chavez, our guide in the Peruvian Amazon
(Photograph by Sy Montgomery)

he begged her, "Please don't tell your family or girlfriends." But she told everyone in Iquitos she would soon be married to a rich man.

"And the dolphin knew," said Moises. "Because the spirit of the dolphin was always watching her."

The next week the stranger returned sad and angry. He told her, "I am very sad. This is the last day I will ever see you." And with that he jumped into the water and disappeared. At that instant, her beautiful shoes turned into big armored catfish. Her ring turned into a leech. Her watch turned into a crab and crawled away. Her necklace became an anaconda.

"She screamed and cried, but didn't want to tell anyone what had happened," Moises said softly. "Three months later, she's in the hospital. She's gone crazy. And the people saw there were hundreds of dolphins in the water by the river.

"My grandmother told me this when we lived in the Napo River with the Yagua Indians," said Moises. "So I know this is true."

Carefully, Moises reviewed our gear for the trip upriver. He found one item conspicuously missing. Even though we both had headlamps and strong little flashlights, he insisted we would need a set of giant flashlights, in case we became lost in the jungle at night. In pursuit of these, we set off together for the market at Belen.

Its stalls are arrayed along the sloping street that spills down to the river and its drowning shantytown. Little booths no wider than a card table offer cigarettes and cassette tapes, socks and hammocks, lace and fishhooks. On the outskirts of Belen, where the average family includes nine children, a month's supply of Lo-Femenal birth control pills goes for the equivalent of 50 cents—the cost of three packs of giant bobby pins from China at an adjoining stall. Mingled with American drugstore fare was the produce of the jungle: piles of dried frogs and fish, the bleached white shells of giant armadillos, the teeth of piranhas, strange fruits with luscious names and sensuous shapes. This large, leguminous inga tastes like a guava, Moises explained; this sapote like an orange. The strawberry-shaped pijayo is the fruit of a palm, whose peach-pit–sized seeds can be roasted like cashews.

The air was thick with the smell of roasting flesh. On charcoal grills, the scaly heads and feet of the river crocodiles called caimans were cooking, and the hocks of peccaries, and shish-kabobs of tiny Andean swifts, their little bills and feet burning black on wooden skewers. One stall sold giant Amazonian snails with shells bigger than a man's fist. A soup made from them, Moises told us, would cure tuberculosis. An old man in the neighboring stall was selling bottles of fermented honey. Moises explained there are ten different kinds of bees in the jungle, but only one produces the honey from which this elixir is made. A shaman must bury the honey in the ground for several weeks, then mix it with various barks and saps; this produces a medicine good for arthritis, and for young women whose wombs are cold. Even in the city, even the most westernized shop owners and hoteliers and restaurateurs still look to the jungle for its magic. There is hardly a businessman in Iquitos who doesn't keep a boa constrictor's head somewhere on the premises, or a jar of dolphin grease, he told us, for it's believed that these items will attract money and women.

In fact, sometimes the charms work too well. Moises told us of a man he knew, named Antonio, who learned from a type of shaman called a *banco* how to make potions and enchant charms. Antonio had collected half a bottle of dolphin fat, and smeared it on his hands; it is said that any girl you touch will then find you irresistible. "I saw this work great," Moises said. Women swirled around Antonio, and finally, the dolphin grease enchanted the most beautiful one of all. Within three months, Antonio made her his wife.

But one day, Antonio went away on business. He returned to find his wife had left him. Moises remembers his friend's anguish: he talked in his sleep, calling for his lost love. Eventually, Antonio sought another shaman's help, and he found out why his wife had left. She had come across the dolphin potion, not knowing what it was. Picking up the greasy jar, she had unwittingly smeared some on her hand. She had been unable to resist the very next man she saw. "You see," Moises said, "the dolphin's power is not always under your control."

Then I felt Dianne's hand on my arm. "Oh—my—God—Sy." She was staring at a short, dark woman in a pink dress. Clinging to her shoulder was a baby pygmy marmoset, the world's tiniest New World primate, a minute, wide-eyed sprite. Its body, not including the ringed black and tawny tail, stretched less than four inches long. It is an infinitesimal, monkey version of a Pekinese: with its large round head, pug nose, and big eyes, a pygmy marmoset is the incarnation of adorable. We had hoped to see one in the wild. They live in groups of seven to nine individuals, communicating with one another in high-pitched trills, warning whistles, and clicking threats. They sleep in tree holes at night. By day, they drink tree sap and with tiny, dexterous, orange hands, capture insects and spiders and pluck fruits and buds. Male and female mate for life and are devoted parents. The father actually assists the mother when she gives birth. He receives and cleans the babies, usually twins, who, like us, are born hairless and helpless. Babies first cling to a parent's belly, and later ride like jockeys on the back. To procure the infant on this woman's shoulder, someone had shot both of its parents. Its sibling had probably also died.

Next to the woman, a black-faced young woolly monkey clutched a man's hair. A titi monkey, head swiveling in alarm, clung to another vendor's arm. I had seen titis at zoos; they sit with their flexible tails intertwined with a mate's or a friend's. At a stall where twenty men were playing cards, one wore a baby white-fronted capuchin monkey like a bracelet. The terrified infant gripped him tight with all the force of its humanlike hands, feet, and prehensile tail, but still screamed with every motion of his captor's wrist. These monkeys had almost certainly been obtained in the same way as the marmoset: the parents had been shot so the babies could be sold here today.

In cages, we found yellow-cheeked parrots, canary-winged parakeets, and more tiny primates, saddleback and black-mantled tamarins —all endangered species. Two baby caimans, two hard-shelled water tortoises, and two more tamarins were packed into a one-by-one-and-a-half-foot-square cage with a baby agouti, an aquatic rodent with a piglike body and a rabbitlike head. The tiny tamarins clutched each other like frightened children. They tore at Dianne's heart. Once, when she had worked as a keeper at the Fresno Zoo, she and her husband, Pepper, had raised two infant tamarins, whose mother for some reason could not nurse them. Both had been charmed by the creatures' agility, their inquisitiveness, and their constant chatter to one another as they raced around the couple's house, climbing the curtains, leaping from lampshades with what surely looked like exuberant joy. Do tamarins feel joy? "I'm sure of it," said Dianne. "Being a tamarin must be a rush."

Atop the cage, another tamarin and a squirrel monkey were tethered by their waists with dirty string. "CITES, Appendix One," Dianne said with gritted teeth. She was speaking of the Convention on International Trade in Endangered Species, which she has repeatedly attended. Yearly, this convention—of which Peru is a signatory—convenes its delegates to decide upon the species of animals that need protection from commercial exploitation. Except for the agouti, the sale of every species we had seen was strictly prohibited by international law. A white POLICIA truck was parked not twenty yards away, but no one seemed concerned; the police do not interfere in such matters. In Iquitos, the officers tend to reserve their attentions for endeavors more lucrative. The desperately rare tamarins were priced at 15 soles each, less than $5; the pygmy marmoset for 20, about $6. I mourned the slain father marmoset who had meticulously cleaned his now-stolen infant, the titi monkey who should be twining his tail around a mate's, the tortoises who might otherwise live for thirty or forty or perhaps even eighty years. But to the people selling these animals, their lives mattered not at all. What mattered to them was the handful of soles.

Feeling sick, we headed back to our hotel. As we left, we passed a woman who was lying on her back on the table of her little stall, her face aglow with rapture. Her hands were occupied and her eyes focused upon something that we couldn't see, because a curtain of leather belts was in the way; we assumed she was playing with a baby. We rounded a bend to see the object of her delight: she was counting coins, dropping one after another upon her belly, savoring each metallic clink. I remembered Moises' story of the boastful girl whose gold watch became a crab. I wished the coins could transform themselves back into animals, and like the crab, crawl away.

Though there are no dolphins in Iquitos's underwater city, we found one in the center of the municipal square.

On the Plaza de Armas, a bright pink, eight-foot fiberglass sculpture springs from the center of a round reflecting pool, perched atop a three-tiered fountain spewing water onto the litter below. The dolphin statue points its snout at a thirty-degree angle toward the sky, its big flippers spread like wings. Dianne observed it looked like an aquatic Peter Pan about to take flight.

The dolphin statue materialized from a dream. Roxanne Kremer, a former mineral trader from California in her late forties, says a vision of a pink dolphin atop the fountain came to her in her sleep one night in 1987. She believes dreams are important: "I set about transforming this dream into reality," she told us. Friends made a mold of the dolphin in her Rosemead backyard, and she had it shipped air-freight to Iquitos in 1988.

The statue is as much a monument to her crusade as to the species. She told us her story when we met her in Miami the year before. Roxanne first encountered pink dolphins on a mineral- and crystal-collecting trip along the Yarapa River, the first tributary of the Marañón. As the dolphins approached her canoe, she said, she began to hyperventilate with excitement. "I saw this pink cloud under the blackish water. There were four or five dolphins around my boat. They looked like a huge, nude human . . . like a mermaid! And they started to communicate with me." She'd been warned not to put her hands in the water, but she couldn't resist: she beat her hands on the side of her canoe, and was rewarded by the dolphins' hissing spouts.

She didn't want to leave them that afternoon, but her guide insisted, warning that the dolphins could come and take her away to their enchanted city if she stayed the night on the river. As she left, she said, a huge rainbow spread itself across the sky; and that night, from her lodge, she heard the coughlike spouting of their breath, as if calling her back. "The dolphins struck a deep chord in me that transcends kinship," she had told us. "They say the dolphins steal you away—and they did. They didn't take me physically, but they did transform the course of my life."

That moment launched a crusade that Roxanne has pursued for the past fifteen years—a crusade to save the pink dolphins.

Roxanne was convinced the pink dolphins were in great danger. Her guide, she said, had told her that people poisoned them, dropped dynamite in their waters, sewed up their blowholes, sold their parts for magic charms. She decided the dolphins needed her help. So she abandoned her mineral business, returned to the Río Yarapa, and founded the International Society for the Preservation of the Tropical Rainforest, whose first project was Preservation of the Amazonian River Dolphin.

She bought a thousand acres of land along the Yarapa and declared it a dolphin preserve. She set up a free health clinic for local people. She organized an International Dedication Day of the Dolphin and Ecology in Iquitos, with marching bands, parades, and art contests for children, still a yearly event. She erected the dolphin statue in the plaza. Today she lobbies for tougher laws and stiffer sentences for animal and skin smugglers. At her Yarapa lodge, she takes in orphaned and injured monkeys and birds—and once, a confiscated jaguar—and there operates an ecotourism venture, Pink Amazon River Dolphin Expeditions, to "raise international awareness" of the dolphins' plight as well as raise money to finance her conservation work.

In Miami, where she was staying with a friend before departing for

Iquitos again, Roxanne bubbled over like the dolphin fountain, outlining her accomplishments and what she intends to do next. She'd like to expand the preserve and her ecotourism operation, hosting scientific and conservation conferences there. She wants to pay guards to crack down on commercial fishermen who seine the river with their huge nets. *International Wildlife* magazine then had a profile on her in press; film companies wanted to make documentaries on her work.

Her dramatic flair, I could see, would make her a charismatic documentary character. She has mastered the art of speaking in sound bites, tells wonderful stories, and dresses for the camera: the day we met her, she was heading to the airport wearing a plunging jungle-print blouse, a tight khaki miniskirt, high-heeled sandals, lots of Peruvian jewelry, and lots of makeup, almost a caricature of the woman jungle explorer. Though she has grown round with the years, she is still an attractive woman. She says the local Indians consider her someone "right out of their legends" of blond-haired goddesses.

But the problem with all this is that I've encountered no one who shares Roxanne's view of the pink dolphins' peril. They are common over an enormous range. Vera and her students have extensively investigated Roxanne's claims that the dolphins are tortured and hunted. Vera stated unequivocally that this is not true. Vera explained that catching a dolphin destroys a valuable fishing net. You will never find dolphin meat in the market; no one eats it. And though the animals' fat, eyes (particularly the left eye), and genitalia are sometimes sold as magic charms to attract money and the opposite sex, they are sold cheap; most of the price comes from the cost of the shaman who casts the spell on the charm, not for the eye itself. The parts sold in the marketplace are, Vera's investigations found, taken only from animals found dead—males who have died of wounds sustained while fighting, animals who have drowned in fishing nets they raid, or dolphins who died from some other cause. Most fishermen are afraid to kill dolphins, for they fear magical retribution.

Though her organization's brochures tout Roxanne's studies of dolphins, scientists we spoke with said they found her fieldwork of questionable value. Roxanne plays music to the dolphins near her camp on an underwater speaker, and then studies the response. "They particularly like Pink Floyd!" she told me. Vera, too, has played various sounds to her study animals. "I played *Magnificat*. I played Rolling Stones. I played 'Tursiops' [the whistles, squeaks, and clicks of the Atlantic bottlenose dolphin]. The response was the same. They are curious about these sounds, so they come to the surface. We should not mix up things," Vera told me. "I like Pink Floyd, but I don't need to assume that they like it."

Roxanne and her work are controversial. Roxanne works from a more "spiritual" perspective than do scientists. Articles about her get published in New Age newsletters like *Earthstar*, the sort of publication scientists often dislike. But even among nonscientists in the animal-protection community, Roxanne has made enemies.

Still, I liked Roxanne; she seemed sincere, and her cause, even if not urgent, was just. Dianne and I both joined her organization for $35, though we were never mailed a newsletter. (Later, Roxanne told me her secretary, a woman she had hired because she felt sorry for her, was suffering from emotional problems and hadn't done the mailings.)

During our meeting in Miami, Roxanne had urged us to visit her at her camp, Dolphin Corners, on the Yarapa, for which she had prepared an impressive brochure, and we considered doing so. But she was not in the country during our visit that spring. Dianne and I promised to stop by her camp anyway and drop off some medical and veterinary supplies with the resident manager.

I found Roxanne's story deeply touching. To finance her project, she had sold her grandfather's land and her grandmother's house in Wisconsin; and in one of our many long phone conversations, she mentioned that she was $56,000 in debt for the project. But the dolphins were more than worth it, she assured: "Dolphins are the highest beings on the planet, for sure," she said. She loves the dolphins; she believes in them; she needs them.

By her own admission, the pink dolphins have seduced her. It may be that, in the shadowy underworld of Iquitos—where, as Jamie said, "you can be anything you want"—she has allowed herself to become the golden-haired goddess of local legend. And the shape-shifting dolphins have let her. She has become the savior of a people and a species—which perhaps do not need her help. Perhaps she is a victim of the same seduction to which so many of the evangelists and aid workers fall prey: allowing yourself to believe that those you love also love and need you back.

Roxanne tells a story that many in Iquitos consider apocryphal. One day when she was swimming in the Río Yarapa, she saw a form in the water that she didn't recognize. She says it was a two-meter bull shark, headed straight toward her. She says that "out of nowhere" a group of pink dolphins appeared and fought it off, while others used their snouts to push her to her boat and safety. "I think," she says, "they were trying to thank me."

The morning we left Iquitos for the jungle, it was raining again: a man rain, Moises said, which would be short and violent, and done in a few hours. I was cold and crabby. I had a sore throat from having been soaked by the rain that had dripped for eight hours down my sleeve through the plastic-covered windows on the *Rápido* boat three days before. (Dianne had chosen the aisle seat so she could get up to smoke in the latrine in the back, perched atop the drum carrying the boat's supply of extra diesel.) As we loaded our gear in our thatched-roof, tarp-shrouded boat, the *Hoatzin*, which would carry us on our day-long journey upriver, I could hardly croak hello to our fellow passengers. Piloting the boat was Mario Huanaquiri, a handsome young man with straight black hair, Indian features, and teeth outlined in goldwork. He'd already married twice, Moises said: "Maybe he has some dolphin grease." Dianne and I christened him "Super Mario," after the video game. Thereafter, we cried out "Super Mario!" with great enthusiasm every time we saw him, upon which he would break into his great golden smile.

Stephen Nordlinger, thirty-one, was a quick, wiry man who worked as a crisis counselor at a Florida clinic. "Your name is Sai?" he asked me with great interest. "That's a kind of an Okinawan fighting staff." Steve, we discovered, is an expert in the Oriental martial arts, who gets together with other guys to fight the way others gather at friends' houses for dinner parties. He had spent the summer between high school and college working as a security guard at an amusement park in Massachusetts. He had slept in the bucket of the Ferris wheel, and by night, dressed in black, patrolled the grounds for prowlers with the Okinawan fighting sticks called nunchaku. His stories were so fantastic that Dianne and I were initially skeptical—until the first time we saw his hand dart out to catch an inch-long beetle he wanted to examine. Faster than a snake can strike, he pinned it with breathtaking precision, its thorax held securely but harmlessly between his thumb and forefinger. From then on, we knew everything he told us was true.

Steve had gone to the Amazon the previous year, with his mother—an intrepid woman who sometimes picks up roadside scorpions and tarantulas on her travels out West and mails them to her son the way other mothers mail their children chocolate chip cookies. On this second trip, he planned to continue his studies and drawings of Amazonian ants, spiders, and insects and to practice his Spanish.

Jerry Goszczycki, forty, was a personal fitness trainer, tall, blond, and muscled. While in Peru, he was going "survival camping" in the jungle. With his guide, Rudy Flores, a twenty-five-year-old with dancing brown eyes and excellent English, Jerry was going to build his own raft to travel down the river and explore the forest. But Jerry was a surprise, too. He was not a macho muscle-head. He was particularly interested in alternative healing, which he was going to study in China on a trip the following fall. His calling card has two weight lifters with his name and address on the front, and on the back:

The Way to Happiness

Keep your heart free from hate, your mind from worry. Live simply, expect little, give much. Fill your life w/ love. Scatter sun-
shine. Forget self, think of others. Do as you would be done by.

Jerry had already had an adventure, he told us. He'd gone dancing at
one of the discos in Iquitos the night before, and there he had met "the
piranha woman." He had only wanted to dance with her, and was
looking for nothing more. They had been having a perfectly fine time,
he said, communicating in snippets of English and Spanish, when
something happened that caused him to end the association: without
warning, she leaned into his neck and bit him.

Eight hours later, as we docked beside Paul's impeccable thatched,
screened lodge, Iquitos seemed a world away. The lodge is built on
stilts, like a tree house, but instead of among branches, this time of year
the building is surrounded by water. The flow of the river follows you

even into sleep. The soft gleam of oil lanterns in the dining hall provides the only artificial light. No generator drowns out the dripping call of the sun grebes, the shivering, rainy trill of the screech owls, the bubbling notes of the tree frogs, the tail-splash of caimans and jumping fish. The ladder-tailed nightjar calls out the name of the place: "Too-WHY-you! Too-WHY-you!" Later, when we would canoe through the flooded forest at night and probe the trees with our huge yellow flashlights, Moises would teach us to recognize the orange glow of the bird's eyes.

The day's travel seemed to have cleansed Iquitos from our city-sore souls. So had the day's rain cleansed the skies. That night, even with a half moon, the stars seemed brighter than I had ever seen. The night skies near the equator are quite different from those we see in North America, with stars I did not recognize. A single constellation sprawled across a full quarter of the sky. I thought it must be Draco, thinking only a dragon could be so big. But in the wet world of the Amazon, during the rainy season, the heavens are ruled not by a fire-breathing reptile but by a water creature: it was Hydra, the water snake. A huge anaconda, many Amazonian tribes believe, is mother of all water creatures. And her allies, I remembered, are the rainbow and the dolphin.

The Rainforest Lodge on the Tahuayo River.

LIFE IN THE RAIN FOREST

Within a few days, I found myself seven stories up a giant machimango tree, hanging by a slender rope over black waters where people were fishing for piranhas. My purpose here was as absurd as my position: I was looking, as usual, for dolphins.

On previous days, with Moises and Mario, we had explored several lakes in our canoe and found that Paul was right: there are many dolphins here. But again, like at the Meeting of the Waters, watching them was like a giant shell game. A piece of a dolphin would appear, then disappear, and we could not predict where it would show up next—or whether the next dolphin we saw was the same or a different individual. From our perch in the canoe, the water's surface seemed like a trapdoor that only the dolphins could open, and which slammed shut with every dive.

Earlier we'd tried to pass through that door at Charro Lake. Less than an hour's journey downriver from camp, past the handsome, stilted village of Buena Vista, the 1,200-acre lake is shaped like a figure 8 with a big mimosa poking its crown up from the middle. The dolphins like this lake in the high water, Moises said, and in fact, in the distance, we saw several right away, rising slow and low. It was safe to swim here, Moises told us—no piranhas. Perhaps, I hoped, I could see the dolphins better if I were underwater, too.

Gratefully, Dianne and I slipped over the side of the canoe and into the cool, wet darkness. The water wrapped around us, weighty and sensuous as a black satin sheet. Then, expelling my breath like a dolphin, I sank wholly beneath the surface. I opened my eyes.

I often swim with open eyes in ponds and lakes at home, and in the sea, and find I see fish and plants quite well; but here the waters were black as night, and I could not see even my hands or feet. Beneath the water, weightless as an astronaut, I felt as if I were floating in a night sky bereft of stars.

The dolphins were still at the other end of the lake, but now we couldn't see them at all. Perhaps they would come over to investigate us. And in fact, after some time, Moises called out to us that dolphins were approaching. Once they came within twenty yards of us. But the visibility was even worse than from the canoe: at eye level, every wave was now as tall as a dorsal ridge, and because our ears were clogged with water, we couldn't even hear their breath. We splashed to attract them; we swam toward them, and away; we floated serene as lilies; but they never came closer.

So we swam, for the pure joy of it. After we had swum for many minutes, Moises said softly, "Don't pee in the water."

"Oh?" asked Dianne, alarmed. As usual, she had been drinking coffee steadily since 4 A.M., and had brought in her voluminous camera bag an enormous Thermos-ful, which she had just emptied. Twice.

"Candiru?" I asked. Candiru are surely the most dreaded fish in the Amazon, far more feared than piranhas. Attracted by urine, these tiny, needle-shaped catfish are known to lodge in the human urethra, where, secured with backward-pointing spines, they are impossible to remove without surgery. One British author, Redmond O'Hanlon, had been so intimidated by the possibility of getting candiru stuck up his penis that before visiting the Amazon he had constructed an elaborate codpiece, which included, as I recall, a kitchen strainer.

Dianne and I, though, were completely safe; candiru do not frequent black-water rivers. At least that's what the books had said. Moises was pulling our leg, as I could tell by his sly smile.

But there were other things in the water, which we could feel but couldn't see. One of them—a fish? an electric eel? a water snake?—had plucked a Band-Aid off the sole of my foot. It could have been a dolphin, for all I knew. For even when we shared their waters, they were invisible to us.

So this was what I was doing up the tree: trying once more to penetrate the black waters. I hoped, at some height, to be able to stare down through the water, to be able to see the dolphins swimming below. I was making slow progress. The climbing apparatus is a harness-type affair, the same gear that mountain climbers use to scale cliffs, with leather loops that grip you around the tops of the thighs and across the pelvis, and a hangman's noose for one foot. You step up into the noose like climbing a stair, and straightening the leg, push yourself up. Then with your hands you slide a knot the length of your step up the rope, and take another step again. At first I was so nervous and clumsy I shook with exertion and fear. But after the first few yards, I felt the treetops calling me up. The machimango was a muscular, sinuous creature, wider at the base than our canoe was long, whose every movement over many decades was now recorded in the architecture of its woody flesh.

Unlike in North American forests, it is easy to see here that plants, like animals, are creatures on the move: You can almost feel the force of limb growth and root grip, the grasp of coiling tendrils, the seductive swellings of fruit and flower. Like animals, plants are hunters: they hunt light. And they are willing to kill to capture their prey. The strangler fig to my left as I climbed had arrived as a seed that became wedged in the crotch of a large tree. Decades ago, the baby plant had

In the trees, life piles upon life: a wasps' nest protects
two neighboring birds' nests.

dropped its aerial roots down the trunk. Fed by the nutrients, the sun-hungry young leaves of the fig had surged skyward, while the plant poured forth more roots like wax dripping from a lit candle. Years ago, the roots had eventually run together, finally entombing the nursery trunk in its clasping loins.

Philodendrons use a different strategy: the seedlings first pursue a tree trunk, and once there, climb to the top by growing modified roots. Eventually, the stem dies at the tree base, severing its connection with the soil. Arriving at the treetop, the plant may wander farther. If its spot is too shady, it can move to a better location by building more

stem, growing in front and dying behind. If the spot becomes too dry, it can drop new roots to the ground. And unlike most epiphytes, many philodendrons do not die if they drop to the ground. They simply begin their climb upward all over again, to perch harmlessly on the host tree without robbing its nutrients.

As I climbed, I willed myself the strength of a philodendron. Up I climbed, past snails clinging to the undersides of leaves; up past an inexplicable frenzy of gnats; up past the big mud termite nest that hung like a clump of drying tobacco in an adjacent, smaller tree; up past a spiky-leafed bromeliad as big around as a Volkswagen. Up I climbed, up past a bird's-foot philodendron with leaves shaped like the outstretched fingers of giant hands, up past the remains of the dreaded wasp's nest . . . up, heave, straighten the leg, pull up the rope, push up the knot. Now cobalt-winged parakeets zoomed by at eye level and yellow butterflies floated below, and still I climbed: up past the grasping tendrils of a liana, up past a brown termite nest larger than a pumpkin, up past another bromeliad in whose teacup-sized bowl I could see mosquito larvae swimming, and up . . .

I finally made it to the top of the rope—up high enough to catch the scent of tiny white epiphytic orchids. They smelled like vanilla.

Now I looked down at the water. It was still and impenetrable as night.

Rain dripped from the trees as we set out for the lake in the morning, to search for dolphins as before. The high water had created new waterways, and with Moises at the bow and Mario at the stern, they threaded our canoe between the tops of low-growing trees. We traveled at eye level to birds' nests and bromeliads, and we could peer into the holes where toucans and parrots were nesting.

Strange lives clung to every tree. White-lined sac-winged bats perched in the lee of leaning snags, angled like tiny bracket fungi. To attract females, the males deposit saliva in special sacs in their armpits and then, once it ferments, they wave their wings to waft the scent to-ward prospective mates. Fist-sized, hairy megalomorph spiders (which we commonly, though incorrectly, call tarantulas) hunt for nestling birds in tree holes. In the way of all spiders, they envenom their prey, liquefying the flesh and then sucking it dry into the stomach, which is located in the head. Every bough, every trunk, every vine around us swelled with life: conical-billed oropendolas flew from grassy, purselike nests hanging from the tips of branches; ants boiled out of huge black excrescences on trunks; brown centipedes curled in the crevices of bark. Vines coiled up trees into minarets, bowers, crenellated chim-neys, and then trailed languidly back into the water.

Every day, we went out to watch the dolphins from our canoe. Each day, we visited a different body of water, even when we returned to the same one: one day we would find Charro Lake had turned as purple as an amethyst; another day it would shine silver. One day at Tibe Lake, where we watched a mother dolphin and calf, the giant lily pads, big as throw rugs, were full of lurid, white, butterscotch-scented blooms crawling with shining beetles; the next time we came, the flowers were gone. On a misty morning at Domingo Lake, a nine-foot-deep seasonal pool that goes dry in the summer, where the dolphins were hunting catfish, we saw a baby perform an unusual, spectacular leap, showing us its whole body. And as we returned to camp, we spotted a spider web that stretched fifteen feet long, perhaps four feet high, draped across the lower branches of several waterlogged águabiche trees. Moi-ses said that perhaps ten fishing spiders had worked together to create it. We had passed this way several mornings before and never saw it; the mist had made it visible. The only things that seemed relatively sta-tic were the sloths. On the way to Domingo Lake, we passed three of them in a mimosa tree, their shaggy fur green with algae that grows in their hair. When Moises whistled, they turned in slow motion to re-ward him with imperial stares. Dianne wanted a photo of one, and hoped that it might move into a better position; but Moises told us, utterly deadpan, "That sloth, she gonna move tomorrow." The next day, there she was with her companions, still in the same tree.

To get to the lakes, Moises took us on shortcuts through the looking-glass world of the flooded forest. We often squeezed between partially submerged trunks of thorn-fringed *Astrocaryum* palms. Even though it was our boat that was moving, the spines seemed to lunge at us. When we saw them coming, Dianne or I would call out, "Spines!" and pull our arms and legs close to the center of the boat. Often just as we were retreating from the spines, one of us would notice that a low branch was also looming at face height and cry, "Branch!" warning the other to duck. Sometimes, our canoe bumped the trunks of the spindly little cecropia trees, which grow like weeds in the high water. Small, aggressive ants rained down on us, into our hair, down our shirts, into our gear, and we would cry, "Ants!" Ants live at the hollow base where the nine-lobed cecropia leaves attach to the stem, and feed on nectar-producing organs on the leaf blades; in exchange for this favor, they protect their host by attacking and biting anything that touches it, as we soon discovered. But Moises was unmoved by our dismay. "But they are not poisonous," he would say, paddling demurely ahead while we picked ants off our skin and threw them overboard, only to find that they could swim and often crawled back into the boat.

The only ants that really commanded Moises' attention were those who injected venom strong enough to disable a person for a minimum of several hours. One of these species inhabits the hollow tangaranga tree, and we quickly learned to identify their host's elongate oval leaves. The giant *Paraponera* ant also garnered Moises' respect: with a glistening black body over an inch long, it possesses a hypodermiclike stinger with such a large venom reserve that it can sting over and over again. The venom is potent enough to incapacitate even a strong man for a day. Then there are ants who bite *and* sting, such as the *Odon-*

A thorn-fringed **Astrocaryum** *palm.*

tomachus ants. They lurk under every log and when disturbed, gush out like blood from a wound, making the sound of crumpling paper when their feet hit dry leaf litter. "This guy," Moises told us, pointing to one of them, "he grab you with his front, then he sting with the back."

Despite the spines, spiders, and ants, the waterways through which Moises threaded us seemed far safer than the land. One afternoon, Moises took the four of us, Dianne, Jerry, Steve, and me, for a hike through the rain forest. At first, every step seemed booby-trapped: we suspected that each log concealed biting ants and poisonous snakes, that trees would drip upon us toxic sap and venomous caterpillars, that

vines would trip us and send us stumbling into palms which would stab us with stinging spines. "Be careful," warned Dianne, as we started our walk in the sweltering heat, "especially everywhere."

Stephen was the best prepared of us four, for he had visited the Amazon the year before with his mother. Sliding down a stream bed, his mother had broken her leg. He was stung by fire ants. A river of army ants invaded their camp and occupied it for three days. Fortunately, Steve likes ants, so this was actually a highlight.

So was the time when a column of army ants hunted him. He had been out in the forest, watching a line of leaf-cutter ants stream by, each holding aloft in its jaws a swatch of leaf like a little flag. These ants comprise thirty-nine species and two genera—all of them gardeners. They harvest small pieces of leaves in order to create giant underground gardens of a fungus found nowhere else on earth—their only food. The amount of vegetation cut from tropical forests by one genus of these ants alone, the *Atta* ants, has been estimated at 12 to 17 percent of all leaf production in the South American tropics where they are found.

Steve had stood transfixed by the flow of these tiny workers, taking photos with his macro lens. "Then out of the corner of my eye, I noticed ants on the left side of the trail, and decided to take a photo of them, too," he said. "And suddenly, the column of them breaks off and heads directly for my feet!

"I turned around and the entire path was closed off with a sheet of army ants," he told us. "I had no choice but to run right through them at full speed. But I was taking pictures the whole time. I kind of figured if I was going to die, someone would find my camera and they'd know what got me." Fortunately, though they did bite him, they never made it up his pants.

Steve had been attacked by another kind of ant, with mandibles as long as its head. It first bit, then stuck its stinger in the wound pierced by the jaws, and vigorously pumped venom from its abdomen into the wound. Chiggers also crawled under his skin. He had even been stung by a *tree.*

Because Steve admires weapons, it didn't surprise me he collected spines. He had found one trunk covered with spines shaped like Hershey's Kisses, and reached for it. Moises had tried to grab his hand, but it was too late: the tree *stung* him. "It injected some toxin that made my hand throb for hours," he said. "I later dissected the spine, and I could never see a mechanism by which it could do this," he said.

"The Amazon is like some nightmare," he said, "where nothing behaves the way it's supposed to." He saw birds fly out of nowhere, perch in front of him, stare him in the face, and fly off. A toad ran—not hopped, he stressed, but ran—away from him faster than he could keep up with it. In the Amazon, he found, one is constantly confronted with realities that simply can't be. Just that morning, canoeing with Graciella, his Spanish tutor, and Juan Salas, the camp handyman, Juan had tried to knock a tarantula off a tree for Steve to collect. It had fallen out of sight, and Steve remembers telling Graciella not to worry: "The spider *can't* be in the boat." But no sooner had he spoken than he saw the giant, hairy legs of the spider crawling up over the gunwales. Steve kept a journal recording his experiences, and toward the end of his first expedition he wrote: "This was a great trip, but now I need a vacation."

But Moises saw a forest in an utterly different way. He had grown up with the Indians along the Napo River, one of the Amazon's tributaries arising from Ecuador and flowing into Peru. He pointed to a vine with reddish bark, as thick around as an anaconda. "This guy," he said, "he give you good survival water when you are lost in the jungle." He whacked it with his machete, and hoisted the severed stem over our heads so its cool, fresh water poured into our mouths. He shoved the severed vine back into the ground, where he told us it would grow.

A few steps later, he sliced a great gray-green tree trunk with his machete. First it oozed red, like blood. But then white sap welled up and dripped like cream down the smooth bark. "This guy," he said, "we call árbol de leche—milk tree. He give you strength in the jungle for two days." The river people, he explained, also use its milk to treat anemia, and the sap can also be boiled to yield a glue to smear on perches to catch birds. On the trunk of another tree grew a little black swelling, like a gall; this Moises lit with a match, and it burned like an oil lantern. A little beetle lives inside the gall, he explained. "This guy," he said, referring to the beetle, causes the tree to make a tar, which the people use to seal their blowguns and burn for light in their houses.

Moises referred to every plant and animal in the forest as "this guy." Where we saw bewildering, erratic growth, Moises saw a forest full of

characters. Like people, some of these characters—the tangaranga ant, or the poisonous bushmaster snake—could harm you. But others, like the red water vine, the milk tree, the copal gall—offer cool water when you are thirsty, food when you are hungry, and light in the darkness.

I remembered what my friend the ethnobotanist Mark Plotkin had lectured to a New York audience years before: "A Westerner looks at the jungle and sees green." But after his first contact with the Indians of the northwest Amazon, he realized, "an Indian looks at the jungle and sees a grocery, a hardware store, a repair shop, and a pharmacy." Mark has spent more than a decade recording the extensive plant knowledge of tribal people like the Tirios, Wayanás, Akuriyos, Waiwais, and Yanomami, and still is learning more.

The people in the villages we passed each day were not Indians, though they had Indian ancestry as well as Portuguese, and sometimes African, blood; but they, too, used many of these plants in their everyday life. From the fronds of the yarina palm, they made their beautiful roofs, which cost them nothing and would last ten years. In minutes, they could weave a backpack out of the fronds of chambira, strong enough to carry home a hundred-pound peccary. They grate their manioc, the tuberous staple of the diet, with the spiny stilt roots of raffia. From a buttress of the great remo caspi, they could slice a canoe paddle—without killing the tree. By twisting the fibers of the leaves of the young *Maximiliana* palm, they could instantly create fishing line. They weave bark cloth from the fiber beneath the bark of the machimango tree.

But perhaps most astonishing, the people also understood the chemical properties of these plants. Tropical plant species, I had learned from Mark, are twice as likely to contain alkaloids as temperate plants. Alkaloids, which often taste bitter, are a product of eons of chemical warfare waged between these plants and their insect predators. But these

Moises drinks from the water vine.

chemical compounds, Mark writes in his wonderful book *Tales of a Shaman's Apprentice*, also "have had a major impact on every culture— if not every person—on the planet." Alkaloid compounds are responsible for the zing of caffeine, the numbing of painkillers, the toxicity of deadly poisons, and the transporting powers of hallucinogens.

The curative powers in these plants sometimes exceed anything medical laboratories can synthesize, Mark says, for their efficacy has already been proven: first by the plants themselves, and then by the people who have used them for millennia. In fact, he points out, tropical plants have already provided some of our most important medicines: the first effective antimalarial drug, quinine, is an alkaloid from the bark of the cinchona tree, first discovered thousands of years ago by Peruvian Indians. Mark firmly believes that the plant knowledge of Amazonian people represents our greatest hope for finding cures to currently incurable diseases, such as AIDS and cancer.

From the yellow-flowering retama, Moises told us, the people here make a tea that cures yellow fever; from the bark of the amasisa tree comes a potion that cleanses the kidneys. The indano, or iodine, tree's orange bark would cure ringworm and heal rashes, and the elephant ear philodendron contains an anesthetic for toothache. The bark of the root of the motello sanango restores male fertility.

If the people here knew the secrets of the local plants better than the chemists at home, surely they must know also the secrets of the dolphins. They were obviously excellent observers of natural history. Perhaps they knew where the dolphins went on their travels; whether they lived in family groups; if they hunted cooperatively; how they cared for their young. I asked Moises and Mario if it was possible for us to visit one of the villages, to learn from the people what they knew about the pink dolphins.

Of course, they told me. In fact, Mario's father and mother lived in San Pedro, only minutes from the lodge. His father, Mario told me through Moises, could certainly tell me about the dolphins, for he knew them well.

. . .

San Pedro on the whitewater Quebrada Blanco is a neat thatched village of 300 people. Most of its residents came here from Chino village,
five or six miles downstream, which began to flood about ten years ago
when the river, as is often its wont, changed its course and began to
swallow the town. Now more and more people move here each year, as
Chino is almost entirely underwater. Piles of yarina fronds lie drying in
yards and on roofs, evidence of new construction.

Mario's handsome, graying father, Juan, and his lively mother, Ilda,
were among the first settlers to build their house on stilts high above
the river. They built the house entirely from the forest: the springy,
two-inch-wide slats of the floor made of cashapona palm, its wide
beams from the red-barked capirona. The house is spacious, and feels
even more so because items not in use are tucked into the rafters:
blowguns and hunting spears; some tin cooking pots; a big drum made
of peccary hide for dances and celebrations; the skin of an ocelot Juan
killed last year because it was raiding his chickens. In one corner, a
three-year-old was quietly playing with a pet woolly monkey. "My
son," said Mario, in a burst of unexpected English, and with a broad
smile. We gave the boy a handful of hard candy we had purchased in
Iquitos. The charming smile he visited on us in return was so like
Mario's I nearly broke out laughing.

I shook hands with Juan. "Please tell him I want to learn about the
bufeo colorado," I asked Moises, using the name by which *Inia* is
known here, which means "the ruddy dolphin." Juan responded in a
mixture of Spanish and Quechua, and Moises translated:

"The people here have a belief that is true," he said. Five or six years
ago, he said, the waters came very high, and dolphins swam beneath
the houses. The parents of young girls told their children not to wash
their clothes in the river in the rainy season. For the male dolphin, he
likes the girls, and this can be dangerous. The male dolphins want to
take the girls away to the Encante; and the female dolphins can come
for the boys.

Last year, Juan said, a dolphin had come for his wife. She had been menstruating (quite a feat for a woman in her sixties, I thought) and left her soiled underclothes on the raft with the washing, which had attracted a dolphin. The next day, she felt a pain in her stomach. The pain, he explained, was the result of an invisible dart, shot from the blowhole of the dolphin. So you must be careful when you hear the breath of the dolphin, he warned; with its loud exhalations, the bufeo can send forth darts more powerful than those shot from a blowgun. You need a particular type of shaman, called a *curandeiro*, to remove them. As part of an elaborate ceremony he must suck the dart from the flesh with his

mouth, and then vomit it out. Some of the dolphins here embody the spirits of witch doctors, he said, and only a shaman can counteract their powers.

"What are some of those powers?" I asked. When he was ten years old, Juan said, he lived in the nearby village of Esperanza. There a dolphin visited his powers on his godmother, Cecelia. Her husband went into the jungle to work, he explained, and sometimes husbands are gone for years. But he came back earlier than Cecelia expected. Every day, he brought her delicious fish to eat; and every night he made passionate love to her. But oddly, he was always gone by the time she woke in the morning.

A year passed. Then one day her husband came home, but did not bring her any fish; he brought meat instead. Why? Because he had never brought her any fish—he had been away for a year!

Learning this, Cecelia became pale and sick. No longer did she want to warm their bed with lovemaking. She grew sicker every day, and wandered about as if in a trance. Her husband could not understand what was wrong. Finally, he sought the help of a shaman.

In a trance, the shaman visited the Encante. Now he saw what had happened: after her husband had left for the forest, a dolphin had fallen in love with Cecelia. But knowing that she would never take a lover, the bufeo had transformed himself into the very likeness of her husband. In his guise, the dolphin had enjoyed Cecelia's favors every night. But there was more to the story, Mario's father said: the dolphin had fathered a baby that was growing in her womb.

In the forest, there grows a vine called tumbo that we had seen on our walk. Twisted, ribbonlike, it reminded me of an umbilical cord. From this vine, women make a potion to induce abortion. Cecelia took the potion, and was delivered of the fetus. "It had a hole in the top of its head," Juan told me through Moises, "the blowhole of a dolphin."

Mario's mother and father, Juan and Ilda Huanaquiri.

. . .

I had been looking for facts about an animal, and found instead stories of a phantasm. I had tried to see beneath the water, and had instead climbed into a plant world in the air. Though still I could not chart their courses through the water, I had followed the dolphins into realms I had never before imagined they might take me—into treetops, inside black waters, through the looking-glass world of the forest's powers. And now, they had led again to new territory: to the people's understanding of the world beneath the river; to the edge of that thin line between animal and human, water and land, fear and desire.

Again, I felt as if I were back up the machimango tree: hanging in midair, staring blindly into the polished surface of some impenetrable mystery. As in Charro Lake, the mystery was still elusive: only now I was surrounded with it, in far deeper than I thought I would go.

DEATH IN THE RAIN FOREST

By now, we had established a morning routine: After shaking our shoes out, evicting giant cockroaches, Dianne and I would wander into the dining hall to check on the movements of the creatures in camp. "Where is the tarantula?" Last night, it had been observed crawling along the porch toward our room, to Dianne's immense dismay. ("What's worse than finding a tarantula in your room?" I had asked her. "Nothing in this universe," she had replied. "Nope," I said. "It's losing a tarantula in your room.") Later, it had retreated to a corner in the kitchen.

"Where is the vampire bat?" Steve had caught it, with his quick, ungloved hands, in Jerry's room, and now it was flying around in the dining hall. "Where is the whip scorpion?" This animal is actually not a scorpion at all but a spider, half again as big as a tarantula, with nine-inch antennae, and folded, hairy front arms to seize insects. "I do not like this whip scorpion *at all*," Dianne had commented. But Steve had released it in the dining hall, and now, he said, "it could be anywhere."

"Where is the poisonous caterpillar?" Steve had collected it from a tree, brushing it with a stick off the leaf it had been eating into one of the plastic containers he always carried for such occasions. The caterpillar has long yellow hairs and white stripes; its venom, Steve said, is as toxic as a coral snake's. He had confined it in a can marked PELI-GROSO but now it was inexplicably gone.

So many creatures here seem armed for Armageddon: They must contend with the fact that at any moment something may be trying to eat you, to strangle you, to sting you or bite you, to suck your blood or lay its eggs in your flesh. Even the great bird-hunting tarantulas had something to fear—the female tarantula wasp. Fully five inches long and purple-black, she flies in search of these huge hairy spiders, in order to sting her victim into paralysis and lay her eggs in its flesh. When the larvae hatch, they feed upon the body of the still-living spider, and when they are old enough, chew their way out of the body.

And yet, there is a strange and exciting beauty to this orgy of hunting and feeding. Each night, in our canoe, we visited a hollow snag that poked up out of the water quite near the lodge like the chimney of a drowned house. We usually came to check on the tarantula who lived there, and we could almost always count on seeing him hunting outside the rim of the hollow. But now, suddenly, he grabbed at something with his forelegs. What was it? The creature raced around the other side of the trunk, like a squirrel. Moises swung the boat around to see. The spider's intended victim was an inch-long orange assassin bug, named for its deadly hunting weapon—a hypodermic proboscis with which it injects its prey with venom. The assassin had just speared a large beetle, but in its haste to escape the tarantula, it scurried away with the beetle impaled on its poisonous spear.

No one appreciated these dramas more than Steve. His stories often began with sentences like, "One day I was out harvesting ants for my scorpion, when . . ." or "The only time I was bitten by a poisonous snake, I'd got this call from my boss at Snake Control. . . ." Outside of Orlando, where he had a home much like our lodge, built over a swamp, Steve had amassed an impressive collection of stinging, biting, poisonous predators. As well as snakes and scorpions, he had a two-foot tegu lizard with a blue tongue, who, he realized, "would love to kill me." The only animal in his menagerie he truly adored was his skunk ("One time when my skunk was combing through my hair, looking for something to eat . . ." began one of his stories). He had bought her because he was afraid of spiders, and skunks love to eat spiders. But, with the same great mental discipline required of a student of the martial arts, Steve had taught himself to enjoy and admire spiders for the very violence he had originally feared.

As he had taught himself the graceful moves of tai chi chuan and karate kata, he taught himself to see the grace in the violence of nature. And this was why he loved the Amazon. It was part of why Dianne and I loved it, too: this vast operatic drama of life and death, where beauty and cruelty twine tight. There is a wholeness to the spider's bite, the assassin bug's poison, the tarantula wasp's sting. None are evil or pointless; rather, the opposite is true. All are fulfilling roles which evolution had taken millions of years to perfect.

But we, of course, were spectators in that drama. For all the biting, stinging creatures here, for all the spines and branches, for all the unseen fish with sharp teeth and poisonous spines, Dianne and I felt comfortable and safe. True, my pale skin had burned badly in the sun; even Dianne's nut-brown California tan was now peeling off her nose like shoe leather. True, we had so many insect bites that scratching them became a form of passive entertainment, like watching TV. I woke up sometimes in bloody sheets from scratching bites in my sleep. But nothing worse had befallen us.

And although we still couldn't identify individuals, we were making some progress in our dolphin observations. One spectacular morning, we had found Charro Lake alive with dolphins—perhaps a dozen bufeos and perhaps as many tucuxis—and recorded 160 surfacings in a single hour. Two bufeos even swam into the shallows, in order to get closer to us. Later we discovered they were swimming in only four feet of water, jammed with submerged branches. Though we'd read bufeos generally ignore tucuxis, that morning the two species were clearly interacting.

Fifteen times in one half-hour period, we saw tucuxis and bufeos in very close proximity, and four times we saw a baby tucuxi surface next to an adult bufeo, the sleek little gray head beside the large, pink, bulbous one.

At one point that day, a pink face erupted from the water bearing a fish sideways in the jaws. The dolphin shook the fish like a dog shaking a sock. In dogs, this is a kill gesture, meant to break the neck of a small prey item; so for the dolphin, the shake may have broken the bones in the fish's body so it could not struggle. The dolphin then repositioned the fish in its jaws so it slid down the throat headfirst.

By timing and numbering our observations, we were able to see some patterns. The bufeos were generally most active for periods that averaged half an hour (although sometimes they stretched to forty-one minutes), followed by periods of rest of roughly the same time. They were most active in the morning; by noon, we often found, they moved away, perhaps to the rivers.

Once, we thought we saw a dolphin sleeping. For eight minutes, between eleven twenty-nine and eleven thirty-seven one morning, in a shallow spot along a grassy bank, we saw an individual rise, slow and low, almost once a minute at the exact same spot. Later we learned that bufeos prefer to sleep in shallow waters, often (disconcertingly, for aquarium patrons) upside down. It is thought, in fact, one reason most of the hundred or so bufeos that had been imported to aquaria between 1950 and 1976 quickly died is that they had no access to shallow waters where they could rise to the surface easily.

We made this observation at a village named Huasi, the busy crossroads of two waterways where bufeos, tucuxis, and people often met. Along with Charro Lake, this was one of our most productive observation areas. We could count on seeing dolphins there almost every day, and here we clearly observed how the hunting techniques of tucuxis

Little girls play with baby crocodiles as if they were Barbie dolls.

and bufeos differ: the tucuxis always hunted in groups, seeming to herd schools of fish toward one another, while the bufeos pursued fish singly.

In addition to the dolphins, children were always coming and going in canoes to the noisy, stilted school. One day, at recess, we saw two little girls holding baby caimans in their hands like Barbie dolls. Bored with their reptilian toys, the bolder girl suggested, *"Vamanos a ver los gringos!"* ("Let's go watch the gringos!") and paddled over to stare at us intently for half an hour as we recorded data on our check sheets. Opposite the school, adults gathered on an island bus stop, awaiting the *collectivo,* the "water bus," to Iquitos. When the boat arrived, again we

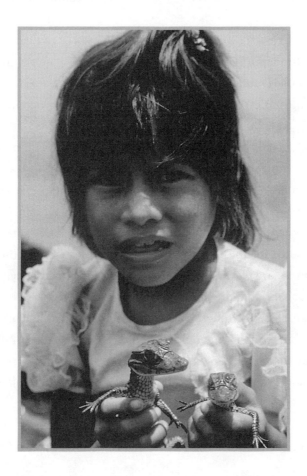

saw a marked difference in the behavior of bufeos and tucuxis: several bufeos came close to investigate, though the tucuxis stayed away. When the water bus pulled away, the bufeos frolicked in the wake and blew air noisily.

With a thick sheaf of observation sheets recording everything we saw, we would return from each outing to Paul's immaculate lodge and feel as if we were coming home. With its good food, cozy beds, cold-water showers pumped from the river, and his attentive staff who had become our friends, the lodge was a cocoon of comfort and safety. We began to feel that nothing bad, nothing pointless or terrible, could ever happen here.

But on the day of our week's anniversary in camp, we saw that we were wrong.

We were headed to Charro Lake that dripping morning. On the way, we passed the *Francis Antonio,* the thrice-weekly *collectivo* from San Pedro to Iquitos. By six-fifteen, the boat's thatched roof was already piled high with logs, bundles of yarina fronds, and twelve lumpy burlap sacks of charcoal. A slaughtered peccary was slung, hammocklike, in the rear of the boat, while most of the passengers sat forward, gazing out the windows. Three roosters perched there, their gold and green tail feathers streaming like wet ribbons in the drizzling rain. The roosters would be roasting by day's end, I thought. I felt sorry for the caimans, sorry for the peccary, sorry for the roosters; but the people, too, of course, are players in the drama of hunting and killing here, and for this they are no more guilty than tarantulas or jaguars.

Next we passed through a canal. Vultures, with their black lizard heads and naked necks, perched watching from the tops of drowning trees, waiting for death to feed them. Moises said dead animals float by this corridor, strangled by vegetation. All life, I began to think, is savage here.

But soon, one of Moises' shortcuts brought us face-to-face with innocents. In the thorny crown of a mimosa protruding two feet above

the water's surface, Moises spotted a cup-shaped nest only slightly
larger than a hummingbird's. Our canoe sideswiped the tree at the very
moment he spotted the nest, and its single, speckled, half-inch egg
popped out into our canoe.

Had we broken the tiny egg?

We heard anguished calls from the edge of the drowning bushes. We
couldn't see them, but we knew the callers were the parents of that
egg. They had seen what had happened. They knew what was at stake:
their entire universe was, at that moment, at risk of flying apart.

Suddenly, the vulnerability and perfection of that egg nearly made
me weep. Frances Hodgson Burnett has written of the "immense, tender, terrible, heart-breaking beauty and solemnity of Eggs." In *The Secret
Garden,* Burnett wrote: "If there had been one person in that garden
who had not known through all his or her innermost being that if an
Egg were taken away or hurt the whole world would whirl round and
crash through space and come to an end—if there had been even one
who did not feel it and act accordingly there could have been no happiness even in that golden springtime air."

Surely the egg was life's first love: Mates come and go, but to the egg,
life has remained steadfast. Love, I thought, may have originated with
the nest—one built perhaps a quarter-billion years ago by one of the
reptilian ancestors of birds and crocodiles, the thecodonts, who may
have guarded eggs and fed their nestlings 185 million years before
Tyrannosaurus rex. Surely, I thought, that reptile brain had known to
fear the destruction of her eggs. Love's twin is fear, fear that the loved
one might be hurt, or taken away. And like love, it is ancient and abid-
ing: it is the architect of the nest, built to protect the beloved. So this
ancient fear is with us still, so old it is inseparable from the most profound and lasting of loves. In this Amazon world where everything
seems at once to be gorging and mating and hunting, where life feeds
routinely on death, still, the parents of eggs and nestlings, parents of
babies and children, feel their anguish no less. A savage world, I realized, is no less loving, no less anguished.

Dianne lifted the egg. Miraculously, it was unbroken. As we replaced the egg in its perfect cradle, we saw the parents: two white-banded antbirds, fat as juncos, with dark feathers pin-striped with white lines. They were hopping anxiously, calling back and forth to one another amid the dense foliage twenty yards from the tree, to which they had fled upon our approach.

Later we passed the large, leaf-lined communal nest of the greater ani, a black, thick-billed bird with prominent white eyes. The nest had been built with the efforts of five or six adults, much like the homes of the people who live here. Several females laid and incubated their eggs together. Surrounded by a cloud of mosquitoes, the babies screamed and the males flicked their paddle tails and growled at us bravely.

The day proved too rainy to observe dolphins, so we retreated to the lodge. We were waiting for Mario's mother to come. When we had met at their spacious thatched home in San Pedro, Ilda had generously promised to show us weaving one day, and she was scheduled to come right after lunch. Dianne and Steve and Jerry and I passed the time pleasantly, leafing by lamplight through the eclectic library of field guides and novels and handbooks. In *Where There Is No Doctor*, a guide written by a health worker in rural Mexico, I read under the heading "Foul or Disgusting Remedies Are Not Likely to Help" that leprosy cannot be cured by a drink made of rotting snakes, nor syphilis cured by eating a vulture. Further, it advised, to cure goiter, don't tie a crab to the lump; don't smear it with the brains of a vulture; do not apply human feces; and do not try to cure it by rubbing it with the hand of a dead child.

By noon, the sky was still fat with rain. The air hung thick as a wet flannel sheet. Mario's mother was late, but we assumed that, in the manner of people who don't live by the clock, she would eventually arrive. Moises and Mario took the canoe to San Pedro to see what was keeping her.

Meanwhile, Jerry showed us exercises to enhance our chi, or life-energy. One has to squat, with the back very straight, knees straight out front—much more difficult than it sounds. With Rudy and Moises, with Steve and Jerry, we debated questions like: what percentage of the time do you think an ant is crawling on you? Our estimates ranged from 100 percent (Jerry) to 10 percent (Steve).

At two-thirty, Moises returned, without Ilda. With a nervous smile, he announced softly: "Some bad luck today." We expected to hear that Mario's mother had confused the date, or had some urgent errand. But no: Mario's three-year-old son, who had been playing so tenderly with the woolly monkey in San Pedro just days before, who had given us the gorgeous smile in return for a handful of candies, had, this morning, fallen into the river and drowned.

He had fallen off the raft docked just outside the house where he had been playing, and was swept away by the current.

Mario, Moises told us, was out in his canoe right now, working with the men of the village in a search party to try to recover the body. Some bodies of drowned people are never found; it is said the bufeo steals them away to the Encante. Actually, the dolphins may eat corpses, for their diet is varied. At high water, chances are better that bodies can be recovered, Moises told us; during this time of plenty, it is less likely that the corpses will be scavenged.

The moment Moises stopped speaking, the rain swept down with renewed force. A woman rain, we thought, that could cry all day.

Steve and I sat stunned. Dianne went outside and cried, then came back and stared vacantly while smoking a cigarette. I sat stupidly repeating, "I can't believe it. I can't believe it." We did not know what to do with the grief, or with the horror, or with the guilt. We had reveled in the danger of the Amazon, thinking we were remote from it. We had observed it like a work of art, like a drama on a stage, not thinking the river, the mother of this place, could swallow one of us whole.

A few minutes later, Moises sat down and in his soft, dreamy voice, announced he had a story for us. Its name, he said, was "Cuento del

Delfín y el Yacuruna." Like a father calming frightened children, he began:

"I remember when I was younger, there was only one village every two or three miles, even near Iquitos," he said. "The tributaries were more quiet. Many different animals lived in the area. But the motors in the Amazon scared the animals. And I think this is the reason many of the phantasms have disappeared." He began to draw, and a face appeared on the piece of paper.

Moises' voice is very low, and he often draws as he talks—often maps of the waterways, to show us exactly where something happened. His words came as softly as the lead strokes of his pencil, the picture and the story taking shape at once.

"This guy here," he said, pointing to his drawing, "he sees the phantasm."

"My grandmother lived in a village in Brazil, which is now a big town. Only two families lived in this town. The nights were quiet. She heard big crocodiles in the night, and dolphins in the river." The nights were more silent then, and darker. The family had no flashlights. For lamp oil, they burned manatee fat in a big bucket.

One day, he explained, when his grandmother was about fourteen years old, her father went to Iquitos to sell bush meat in the market—a week-and-a-half-long trip by canoe. She and her brother were alone in the house, with a canoe, a shotgun, and two dogs.

Their fourth night alone was a moonless one. She and her brother sat on the front of the house by the wide, rain-swollen river. The manatee oil burned in the corner. She looked at the river and watched it move. A dolphin gasped.

And then, perhaps a mile downriver, they saw a great soft light shining from beneath the water—shining like a diamond. They asked each other: what could it be? They had never seen light like that. Watching, transfixed, they saw it was a great boat, bigger than a house, coming toward them slowly, almost magisterially, and incredibly, making no sound.

The magnificent boat came closer. Finally, it was close enough for her to see inside: It was packed with people, and the sounds of their laughter and music drifted to the children on shore. Men were dressed in wide hats, black shoes, white pants, and there was a beautiful girl dressed like a princess. In the wheelhouse of the boat, she saw a big man reclining like a king. His teeth were pure gold, and the diamond-light, she saw, was coming from his mouth.

The two dogs began to bark, and the young girl who would become Moises' grandmother grew scared. "Take the gun!" she shouted to her brother. He fired three shots.

The boat began to sink slowly back down into the water. But the diamond-light strangely rose. Within ten minutes, the boat had vanished—but in its place, pink dolphins leapt and blew.

The next night, Moises' grandmother and her brother went to stay with the other family in the village. But on their way, dolphins surrounded their canoe, as if they were trying to convey some message.

The message, Moises explained, came to his grandmother in a dream. The Yacuruna said to her, "Why were you scared when you saw my boat? I am a rich man. The beautiful girl you saw is my wife, the princess. I have many employees in the boat. The people in the white clothes you saw, they are dolphins. I have a big city in the water, with towers and a palace. I want to give you diamonds and gold, if you will come to live with me."

When her father came home, she told him about the boat, the light, and the dream. But he didn't believe it—not until later, when he went fishing in a canal near the mouth of the big lake.

He was looking for tiger catfish, which are best caught at night, and had brought along his cigarettes and his harpoon and long lines with big hooks. In the middle of the night, he felt something big pull on the line, and he pulled hard, but it wouldn't give. It kept pulling, so hard that his lamp fell into the water. Finally, he felt his canoe might go underwater, too, so he hurriedly tied it to a tree. Later he returned with

his son, and the creature on the line again pulled so hard that the canoe seemed to be fighting huge waves. "Something is trying to catch
me!" said the father. And the son said, "I know. I saw his boat."

After that, the father fell ill. The son had to fetch a shaman from
Tabatinga in Brazil—the best shamans used to live there, Moises explained. A learned shaman was found, and he sought the answer in a
trance. In his dream, he saw the Yacuruna, and spoke with him, and
once the trance was done, the shaman revealed what he had learned.
"You are a good man," the shaman told the father, "and the Yacuruna
wants your daughter to live with him." The shaman blew tobacco
smoke on him and sucked on his stomach, and the illness was gone.
But the Yacuruna was still there, and he is there still. There was nothing the family could do but move to another town. They moved to a
village called Malupa.

"But what if your grandmother had gone to live with the Yacuruna
and the dolphins underwater?" I wanted to know. "What is it like
there?"

"The shamans say life in the water is the same as here, but better,"
Moises said. A shaman in Iquitos had told him, for instance, that underwater there are more hospitals. There are epidemics here—malaria,
cholera—but not in the Encante. People live longer there; one month
underwater is the same as a year here. But even in the Encante, there
are dangers. The shaman told Moises the story of a young man who
went to live beneath the water.

One time, at the mouth of the Napo River, a young man's boat disappeared, and his family was devastated. But just a little while later, the
young man appeared to his mother in a dream. "All is well," he told his
mother. "I am married to a beautiful girl who is a princess. I live underwater, in a beautiful city. It is good here," he said, "but tell my sister not
to wash her clothes in the river, or the dolphins may come to steal her
away to the Encante."

For twenty years, he lived happily beneath the water. But one day,
he told his wife, the princess, that he would like to visit his mother and

sister on the land. His wife told him, "Until you return to me, I will give you a little rock, and it is your life. Please don't lose it." So the young man came to visit his mother and sister. As the mother was taking water from the river, she saw her son coming toward her in his canoe. They hugged and cried.

He kept the stone in his pocket, and when he swam in the river he could touch his wife. But one night the village held a celebration, with masato and liquor, and he lost the stone. The following morning, his mother found him dead in his hammock. His hair had turned white, like an old man's.

The story ended, and we sat together in the lamplight, thinking of death. Then suddenly, a terrified scream erupted from the direction of the cold-water showers—a woman's voice, Graciella or Gladys. Without thinking to grab my flashlight, I ran toward the voice. About halfway there, I realized I was running in the dark toward something that made a person who lives here scream. Gladys had collapsed near the sink and was crying hysterically. *"Onde fica perigo?"* ("Where is the danger?") I asked, forgetting that Gladys speaks Spanish, not Portuguese. *"Serpiente,"* she said between sobs, and then for some reason I demanded in Bengali, *"Shap kothai?"* ("Where is the snake?") and in English, "Did it bite you?"

A contingent of other staff as well as Dianne and Steve appeared in the dark, and in a second, with his lightning hand, Steve had caught the snake, pinning its head gently but firmly between his right thumb and forefinger. It was a three-foot yellow tree boa, whose triangular head gave it the look of a viper, but it was harmless. Steve handed it to me and let it wrap its lovely length around my arm, like a caress. I felt immense gratitude toward this snake: I was grateful that it had not bitten Gladys, that it was not poisonous, that I could touch it without fear. At that moment, I fervently loved that snake, for I could admire its beauty without fearing its cruelty. Unlike the Amazon that had swallowed Mario's son, I could adore it wholly, and feel no guilt for loving it.

We released the snake by the front steps leading to the dock, where we expected it would reappear on a fishing pole or in our boat the next morning.

After dinner, Moises had an announcement: the little boy's body had been found, and there would be a gathering at the house that night. We were all invited.

Dianne felt we shouldn't go. She and her husband, Pepper, knew what it was like to lose a son. At the time, she hadn't wanted to see anybody. She had hated having to smile at the well-wishers, who would say things like "I know how you feel," when they actually had no idea, and counsel, "Time heals all wounds," when nothing ever heals such heartbreak, not really. What would Mario's family want with a bunch of rich tourists gawking at their dead son? She felt we should butt out.

But perhaps, I worried, Mario's family would be offended if we didn't go. I asked Moises what he thought we should do. The family, he said in his soft voice, would consider it an honor if we attended.

The front of the house was jammed with canoes when we arrived that night. Ours was the only boat with a motor. Five men were sitting outside, smoking, atop an overturned canoe. The grandfather, Juan, whom we had met here just three days ago, greeted us with a smile and a handshake and *"Buenas noches."* We stood silently outside for a few minutes, then mounted the steps to the stilt house and joined the crowd inside.

About fifty people were there. In the flickering lamplight, we could see men playing casino, gambling for cigarettes, in a corner behind the mosquito net. Women lined one wall, sitting on wooden benches and rocking children in two hammocks while piles of other children slept like puppies at their feet. It seemed like a party, except the peccary-hide drum was still in the rafters, stilled, and no one was dancing. But no one was crying, either. In fact, people were laughing and telling stories. Everyone, including the few children who were still awake, was drinking masato, the beer made from manioc root, which the women chew and spit out to ferment in a jar, or, for a really big party, in a dugout canoe. The grandfather, Juan, offered cigarettes to everybody with a kind smile, exactly like a hostess passing canapés at a party.

In the midst of it all, Mario's son lay dead on a table, his little body wrapped in white cloth decorated with red plastic flowers, a crucifix at his head and two candles burning at his feet. Mucus bubbled from his nose periodically, and his older sister, perhaps five, unceremoniously wiped it with a handkerchief, a motion she had completed hundreds of times before. She had done this for him while he was alive; why should she not do so now that he was dead?

Soon Mario appeared, wearing a towel around his neck, joined by a friend wearing a T-shirt that proclaimed DO THE WILD THING. "We're so sorry, Mario," we said in English, reaching for his hand. Mario smiled at all of us, his great, gold-glinting smile. He was glad we had come.

Ocelot skin, a common household furnishing.

Then Mario turned to his son on the table. Together with his friend, Mario slid out a tape measure alongside the little boy's body, running from head to foot. Minutes later, we heard a handsaw at work. They were making the coffin.

We sat down on benches and looked around the room. Almost every woman of fertile age was visibly pregnant, including one gray-haired, stooped grandmother who wore her pregnancy like a basket too heavy to carry. Imagine having to bear a child at her age, I said. "I would shoot myself," said Dianne. Yet the abortion vine is growing everywhere, I observed. The crucifix at the head of the child's bier, though, probably put an end to much of that.

I wondered how these pregnant women felt, here at the wake of their friends' child? My friends at home would have felt guilty, I knew, thinking the great ripe fruits of their bellies an affront to the bereaved parents. I remembered how one friend of mine, who had just given birth to a healthy son, had grown shy around a woman friend whose baby girl had been born with a slight deformity; another friend, newly pregnant, would not discuss her pregnancy with a girlfriend who was barren. Both felt guilty for their own fortune. But this is not the way of the people here. Death is no freak in the Amazon, but a companion with whom one walks daily—not without fear, not without mourning, but with composure and grace.

Most of the people in this house would stay the night with the family. Some would sleep, and some would play cards and smoke and drink masato and tell stories. They offered their presence like a gift, to stay beside the family to prove they aren't alone. They don't defy death, but they defy loneliness.

Like people confronting death everywhere, everyone brought food: Food equals life. One person brought a rooster. And just before we left, we saw Mario carrying it by the feet, a tin pot in his other hand, in the direction of the kitchen. Then we heard it scream.

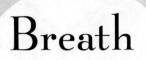

Breath

ON THE GLASS-SMOOTH WATERS OF CHARRO LAKE, WE WAIT for the moon.

A bat shoots by, a flying shadow. In the dark, frogs call in the clicking voices of bamboo chimes; others twang like rubber bands. Lightning pulses silently in the southern sky. But the moon hides behind dark billowing rain clouds, so we wait.

Patron and seducer of women, the moon opens women to nighttime lovemaking, Amazon myths tell us; it controls the cycles of plants and menstruation. The Shipibo say that by moonlight, women are especially susceptible to spells cast by dolphins. The moon, it is said, is the sun of the underworld. It traverses the world of darkness, illuminating the Encante. That is why Dianne and I have come here now: with Moises and Graciella, with Jerry and Steve, we wait in our canoe for the moon to reveal the dolphins.

At 6:40, a glow pierces the clouds. By 6:45, it shines bright enough to write these words by. And still we wait.

Around us, bells, creaks, whistles, honks; the forest heaving and sighing, like a dream set to music. And then, at 7:04, we hear their breath: "Chaaahhhh!" A minute later, another gasp.

Another minute passes. And now, all around us, tiny bubbles begin to rise—behind us, in front of us, to starboard, then port. It is the expelled breath of dolphins—breath so close we can touch it.

I stop taking data. "This can't be happening," I say to Dianne. "Believe it—it's happening," she answers. I dip my hand in the water and feel the bubbles sizzle on my skin, intimate as a caress. It seems as unreal as a kiss from a ghost, and yet it continues: for four minutes, we touch the intangible and see the invisible, as the dolphins bless our canoe with their breath.

This communion is a magic that we, in the modern West, have long forgotten, but one humankind once knew well. The ancient Greek word *psyche,* which signified "soul" or "mind," also meant "breath" or "gust of wind," and was derived from the verb *psychein,* which meant "to blow." Another Greek word for "air, wind, and breath"—*pneuma*—also signified what we call, in English, "spirit," as in the Latin word *spiritus,* which gives us dual-meaning words like "inspiration." The ancient Sanskrit word *atman,* which means "soul," also means "air," as well as "breath." But as David Abram points out in *The Spell of the Sensuous,* breath, air, and soul are the most primal and meaningful of connections: we are all, he writes, animated (from the Latin *anima,* also meaning "breath, air, and soul") by the same currents, by the "living, sensuous, holy air." This can't be happening! I am thinking as the bubbles sizzle around us—for breath is air and air is absence, invisible; "The most outrageous absence known to this body," writes Abram, ". . . the air can never be opened for our eyes, never made manifest." But the Navajo, he notes, can see it: They say the whorls at the tips of our fingers are where the animating winds of life leave their trace. The wind-whorls on our toes hold us to Earth, and those on our fingertips hold us to Sky; this is why, they say, we do not fall when we move about. It is breath that anchors us to life, and soul that anchors us to body.

The bubbles disappear. In the seventeen minutes that follow, we can see, by moonlight, the glistening pink heads bobbing in the water. But still, more real than the visible are these breaths: a loud blow, a tail slap, a sigh; a gasp, a blow, a ballet of breath. And now the breaths grow fainter, more distant, as the dolphins move away.

As we are leaving, Graciella, who grew up in a village, says she sometimes sees little stars slowly moving when the moon is out. There—can we see it? Dianne and I tell her it's a satellite, but our answer makes no sense.

VINE OF THE SOUL

"No rain last night."

My first words to Dianne each morning were usually about the rain: what time it had begun, how long it had lasted, how it might affect our dolphin observations that day. The rain had become like a companion to us, whose daily activities were as important as our own. Many days had passed since we had attended the wake at San Pedro, and almost all of them had brought rain. But today we were thirsty for it. There had been no rain the day before, nor that night. I had missed its presence even in my sleep; my dreams had been dry and strange, like an orchestra with no strings.

"The water is going down," Moises told us at breakfast that morning. Already, the water level at camp had dropped two inches. The fish, he said, would go to the Amazon, and the dolphins would follow.

So, crying "Spines! Branch! Ants!" we threaded our way north through the flooded forest to one of the channels leading to the Amazon—thick, brown, and wide as a sea. We waited, but saw no dolphins there—not even Moises, with his Amazon eyes. So Moises directed our canoe through a young, mostly submerged forest of cecropia (as our cries of "Ants!" increased) to a little thatched stilt house by a flooded breadfruit tree laden with green globes the size of grapefruit. A beautiful young woman in a blue dress was swinging her baby on the porch.

"Where have all the dolphins gone?" Moises asked her in Quechua.

A long conversation ensued, with much waving of arms. Later he told us they had all gone to a lake with no name, an hour from here. We did not have enough fuel to get there today. But that night, we would meet the dolphins another way.

We had arranged for Ricardo Pipa, the local shaman, to come to camp. Ricardo, sixty-two, is a thin man with a wide, humble smile, a close-cropped mustache, and a deep cough. Quite soon after we had arrived in Peru, Dianne and I had met him at a logging camp named Ramirez near the lodge. While Dianne played with a captive squirrel monkey, I had talked with him about the dolphin spirits, as Moises translated.

Although Ricardo was gracious and Moises worked hard, it was one of the most difficult interviews I had ever conducted. My hearing is bad to start with, the victim of too many old boat motors on Third World rivers, and Moises' voice is so soft I often couldn't tell when his Spanish had ended and his English translation had begun. And on this day, his English was flagging: he mixed up "man" and "woman," "he" and "she," "week" and "month," and he dropped words as casually as one peels a fruit. To clarify, I sometimes asked him to spell a word, making things worse: in Moises' alphabet (which I was only able to decode much later), what is pronounced as *A* means *E* and *E* means *I*. In addition, while we were talking, we were under attack from a plague of biting triangular horse flies swarming up from the water beneath the slatted floor on which we sat. As they bit our thighs and buttocks we slapped at them violently, derailing Ricardo's narrative. Also, both Ricardo and I were afflicted with coughing fits ("Jungle cough. Gone in a month," Moises had assured me). Later, when I played back the tape of our interview, Dianne and I howled with laughter: each of my questions in English elicited a series of mumbles in Spanish and Quechua, violent slapping noises followed by curses in three languages, and periodic coughing duets.

But I was able to glean how, twenty-five years ago, Ricardo had learned the powers of the dolphins.

He was already studying to be a shaman at that time, he said, having learned the secrets of many plants. He went fishing one night on the Amazon, when he heard a voice calling him from the dark forest. He followed the voice to a tall, pale woman with long blond hair. Together they walked to the water's edge, where he saw a rock that shone with a light as brilliant as the moon. He picked it up, and as he did, he saw many dolphins jumping in the water. This light, the woman said, is my *encanto*—the power that I want to give to you. And then she disappeared.

But she came to him in dreams. Together they traveled with the dolphins to the great underwater cities of the Encante. She taught him the *icaros*, the prayers to call the dolphins. For six months after that, dolphins would follow him by day in the water, and at night in his dreams. The dolphins showed him their powers. They are the protectors of all the animals in the water, he said—the fishes, the caimans, the stingrays, the boas. "The dolphin doesn't like it when people fish too much," Moises interpreted, "and then he breaks the nets." Moises here added that he knew this was true because, once, an older brother had caught a big fish with a thrown bamboo harpoon. When the fish became tired, a dolphin came and pushed the fish away from his brother's canoe, and broke the harpoon in half. "And wild pigs, when they cross the river, the dolphin is angry. The dolphin can kill those animals. He is boss for the water."

The dolphins also taught Ricardo their medicines, their prayers and chants, he explained. It was the dolphins who taught him to speak the Huambisa dialect, he said, and they also gave him many *icaros*. While he was learning from the dolphins, he was forbidden to hunt land animals. But the dolphins brought him great luck in fishing.

Today he still has the *encanto* the beautiful woman gave him. Where is it? He doesn't keep it in his house, Moises explained; he keeps it in his mind. Still, many different witch doctors try to steal it from him; catfishes and caimans, too, try to steal it away. But the dolphins always protect him; and he can call the dolphin spirits whenever he needs them. He calls them in a trance, which he achieves by drinking a powerful potion called Ayahuasca, made from two of the most sacred plants in the forest, toé and ayahuasca.

I had heard of Ayahuasca. In Quechua, it means "Vine of the Soul." Many Amazonian tribes use this powerful hallucinogen to achieve trances in their sacred rituals: The Jívaro of Ecuador call it yagé; the Tukano of Brazil call it caapi. The great English botanist and explorer Richard Spruce recorded reports of the drug's powerful effects: "The sight is disturbed and visions pass rapidly before the eyes, wherein everything gorgeous and magnificent they have heard or read of seems combined. . . ." But, his report continues, "soon the scene changes; they see savage beasts preparing to seize them, they can no longer hold themselves up, but fall to the ground."

Moises had taken Ayahuasca himself thrice—and he did not plan to do so again. "In ten minutes, you don't feel your bones in your body," he told us. "It comes too much in your head—bad things, angry things." The visions can be terrifying—in one, Moises had been attacked by vampire bats. Almost everyone who takes the drug vomits copiously, and yet the nausea persists long after the stomach is empty. Some people shriek in uncontrollable fear. Most lose control of the bladder and bowels. But shamans are different, Moises explained. Ricardo, he assured us, is a very strong man, and he uses Ayahuasca several times a month. He has learned not to vomit the drug, for if you do, he said, the Mother of the Vine leaves you, and can no longer lead you to the spirits of the underwater world.

Before we left Ramirez, Ricardo and I had come to an agreement: the night before we were to leave for Roxanne Kremer's camp on the Yarapa River, Ricardo would come to the lodge and perform a ceremony.

With the help of the ayahuasca vine, he would ask the dolphin spirits to come to us this very night.

. . .

The ceremony began at eight-thirty in the room next to ours at Paul's camp. Ceremonies are usually conducted at night, Ricardo explained, because he can talk with the spirits more easily when there is less noise. We sat in a circle: I at Ricardo's left; Dianne to my left; Moises to her left. A hurricane lamp glowed in the middle of the floor.

On an earlier night, for a good-luck ceremony to bid Jerry farewell, Ricardo had worn a headdress of blue and gold macaw feathers and an X-shaped halter made of red beans, bamboo seeds, and porcupine quills "to attract the spirits." That ceremony was conducted by lamplight, and Ricardo told us that we could speak and even take flash photos. But tonight, Ricardo wore only brown pants and a short-sleeved shirt of white cotton. Once the lamp was snuffed, we should not turn on our flashlights and we should not speak. The Mother of the Vine, not Ricardo's finery, would attract the most powerful of spirits, and we should be careful not to disturb them.

Ricardo lit his pipe. It was not filled with tobacco, but with a type of incense made from a plant in the same family as myrrh. He blew the honey-sweet smoke over the small aluminum pot of Ayahuasca—"to put in the *icaro*," explained Moises—and then drank three deep draughts. He stood up briefly to go outside and spit over the rail.

When he returned, he began to shake a bundle of shaka palm leaves to call the spirits. He blew smoke hard and close on the top of my head—where a dolphin would breathe, I thought—and shook the palm over each of us in turn. I was first, and the blessing showered my head with little gnats who had sheltered in the leaves earlier while the bundle had sat on a table. Ricardo snuffed the light, and he began to whistle in the darkness.

The tune seemed as innocent as something a schoolboy might whistle on a summer day; the whistling changed to a chant. The words sounded to me exactly like the Onondaga language I had heard when I was in college, when I would visit the reservation near the university and feed apples to the buffalo. The words I found calming, and sweet as incense. Beneath my closed eyelids, I saw caterpillars, then butterflies, then leaves, then birds, a gentle stepladder of becoming. I thought: What a kind and gentle man to shepherd us to the Encante.

Time stopped during the ceremony. The sound of the shaking leaves reminded me of rain; between the cadence of the leaves and the mesmerizing whistling and chant, I felt myself slipping toward sleep. Sleep, I decided, is a kind of water, and dreaming, a kind of river. I scratched my insect bites to stay awake. I felt an unusual peace slip over me.

And then, hours or minutes later, we felt a change: The close, hot room suddenly was refreshed. The leaves now sounded like a jar of water being shaken, and I realized it had in fact begun to rain. A word came into my mind: mother. The rain came harder and gathered power, and Ricardo began to speak in a different language. I reached out and held Dianne's hand.

Thunder and lightning concluded the ceremony. We crept out of the room, leaving Ricardo to rest. Moises interpreted for us what had happened:

Ricardo had spoken with the Mother of the Vine. He had met with the dolphin spirits. And yes, they were willing to speak with us.

"They want to come to you," Moises translated. "The dolphins want to teach you their powers. And when dolphins teach you that power, the spirit will be with you everywhere, even in the States, all the time.

"They want to come to you," he said again, "and they will speak to you in English, in French, in Spanish, and in Quechua. They will speak directly to you.

"Tonight," he promised, "they will speak to you in your dreams. Tomorrow, Ricardo, he explain the dream. The dolphins will send dreams for three nights. Ricardo will tell what they mean. But if you learn the secrets of the vine, they will speak directly to you, in voices you understand.

"The rain is good for concentration," we are told.

"The dolphin wants to talk with you very long," Moises said, "and if you want to see the dolphin city, the Encante, he will show you. He will take you there."

But to do so, he said, we would have to drink the vine.

As Moises told us this, at each pause, there was thunder or lightning. "The rain is good for concentration," Ricardo had told him.

It rained all night. At 6 A.M., when Dianne woke me, it was still raining. It seemed auspicious, she said, that the Mother would send a woman rain to us two women, seeking knowledge about the dolphins.

I woke and tried to recall my dream. It seemed disappointing—too simple and vague to be a vision. It was only this: Dianne and Moises, Mario and I, took the canoe to Ricardo's house. We had a large egg-shaped object to hand to him—the shape, I now realized, of a beautiful, light green egg case of a snail we had found the day before. I handed it to him and he took it. There were no words to the dream, no other people, no other objects. The meaning of the dream seemed obvious to

me: Clearly, I was dreaming of what I hoped to do the following morning. The egg case, with its seeds of new beginnings, was my dream, and I was handing it over to Ricardo.

Dianne's dream was dramatic, long, vivid, and complex: it involved a handsome suitor, an old man, wedding rings, coffins, and wailing mourners. Obviously, my dream was inferior. I was sullen at breakfast, poorly concealing my jealousy.

Ricardo had stayed the night in our camp, and he had agreed to come with us to Huasi, on the way to his house at Esperanza, to watch the dolphins with us, and to interpret our dreams. Mario tucked the boat into a grassy bank. Moises asked me for my dream first. He discussed it with Ricardo, and then replied: "In your dream, you want to see the dolphin spirit, but you will not recognize him. He maybe appear as your friend, your husband or wife. Many things in dreams can be transformed."

The canoe in which we traveled in my dream, he explained, was the dolphin's boat—the boat in which he would take us to the Encante. And the giant egg case? "When Ricardo blew on your head, he asked the Mother of the Vine and the dolphin spirit to give you power—a shell, a gun, a spear, a machete. The thing in the boat is the thing the dolphin wants to give you.

"When you drink the vine, the dolphin spirit will give you a defense also. When you drink the vine, you'll receive big powers," Ricardo promised through Moises. "Jaguars and eagles, big demons and big bats—they will help you, and stay behind you always."

Now Dianne came forward in the canoe to reveal her dream: "I had a handsome young lover, but there was also an old man in love with me," she began, and my heart sank lower. In her dreams, beautiful Dianne has two lovers vying for her; my dreams deliver only the egg case of a snail.

"I was not in love with him," Dianne continued, "but I chose him over the young lover. . . ."

With the old man, Dianne went to the jewelry store to pick out a wedding ring. The jeweler showed her case after case of big diamonds. The old man, smiling, urged Dianne to pick the biggest one; but she said she would be happy with the smallest. Then Ricardo appeared in the dream, and he handed her another box filled with diamonds. They went into an adjoining room to look at the rings, and found the room was full of beds with dead people lying in them. When Dianne picked out a ring, it turned into yellow plastic with a gray gorilla on it. How odd! In the dream, Ricardo told her not to worry, that she should wear the plastic ring till the diamond ring was ready.

Then Dianne and the man walked outside. Beneath the store, they saw people dressed in black, wailing. "But it was OK, because we were very happy," Dianne said. "And then the old man walked into the jungle."

Ricardo was pleased with the dream.

"The black young man is a gray dolphin," he said, meaning the tucuxi. "The tall white man who give you the ring is the pink dolphin." Both these interpretations surprised us because in Dianne's dream, the young lover had not been black, nor the old man tall. They may well have become so in Moises' interpretation.

Ricardo continued: "The ring is the dolphin's powers. The yellow color is good luck. Someday, if you get sick, the gold color will help you. The diamond is your defense. Glass is good luck for you. So is clear water.

"The rooms the dolphin shows you are his home. The beds are the beds in the hospital." (The Encante has very good hospitals, Moises had stressed to us earlier.) And the wailing mourners? This, too, was a good sign, said Ricardo—these were really dolphins praying, and rejoicing at seeing Dianne.

"Your dream is very good," he said. "All the people you see are dolphins. The dolphins want to give you many things. The dolphins looked last night more for Dianne than for Sy."

I felt an utter and complete failure: I could not follow the dolphins in the water; I could not spot them from the treetops; and now they eluded me even in my dreams. I despised myself with a vehemence that took me by surprise. It made me wonder whether the antimalarial drug I had been taking for a month and a half was making me crazy. Later, Dianne and I would discover that this is in fact a known side effect of the drug.

Dear Moises must have seen the disappointment on my face, for he added, gently, "Sy, you will have a dream the next time."

For the next two nights, he said, we should record our dreams, and Ricardo would interpret them for us. We would be traveling to Roxanne Kremer's camp on the Yarapa then, bringing her camp manager the supplies we had promised. When we came back to the Tahuayo, though, the spirits would speak directly to us; for on that night, with Ricardo there to guide us on the journey, Dianne and I would take the vine.

The journey to Roxanne's camp took nine hours. Within fifty minutes we joined the Amazon, or Marañón—"The Mother River of the Amazon," Moises explained, whose headwaters are born in the cloud forest. From the Amazon, we would take the Río Ucayali, whose tributaries arise from Machu Picchu, the ancient Inca city in the Andes, east via small channels to the Yarapa, to the lodge Roxanne named Dolphin Corners.

The Amazon was wavy like the ancients' wine-dark sea—red, even though it is technically white-water here—and can stretch a mile wide. The opposite bank—with its tall stands of caña brava grass studded with dangling tassels like the barbed tips of spears (and in fact, Moises said, this bamboo is used to make spear shafts)—seemed far away. But it was not a bank at all, but one of thousands of islands that rise and fall here like waves, as the river, snakelike, eats and digests and finally excretes the land. "When I worked here three years ago," Moises said, "the Yarapa was parallel to the Ucayali." Now they join at an angle. At the mouth of the Yarapa, he told us, there is a sunken village. Three

years ago, he watched the river swallow it. The entire village disappeared in a month. "My father said to me," Moises recalled, "that water
is a natural thing. Man don't stop that thing. Just the God."

And so it should be, Dianne and I nodded in agreement. But this is
no longer true. We remembered that one of Vera's chief fears for the
dolphins was the hydroelectric projects Brazil envisioned for its river
systems: a series of some eighty dams were proposed in the 1980s to
harness 5,828 megawatts of the Amazon's yearly energy for southern
Brazil. Each dam blocks the routes of the migratory fishes, she pointed
out; one small dam, in the state of Pará, caused a 77 percent extinction
of fish species. Only five species are left there, and only two of those are
still abundant. The dolphins, unable to adapt from a diet of over fifty
species to only two, are nearly gone.

We soon saw more evidence of man's growing power over the Amazon—a tugboat pushing two giant barges loaded with a herd of yellow
CAT bulldozers, massed like mechanical dragons. Moises said they were
headed for the Pastaza River, where American companies were drilling
for oil. The Pastaza arises from Ecuador. In the 1930s, Royal Dutch
Shell discovered pools of floating oil near the Andes's eastern foothills,
and by 1938, the company had blasted a route along the cliffs of the
Pastaza Canyon—the first road across the Andes in Ecuador—and the
beginning of a 315-mile pipeline that traverses the Andes and oil drills
that spread across the forests of Ecuador and into Peru. Most of
Ecuador's oil production, some 300,000 barrels a day, comes from the
Amazon region, and its waterways suffer. In July 1992, the month after
the Earth Summit Conference was celebrated in Rio de Janeiro, Brazil,
a major oil spill in Ecuador released 27,000 gallons of oil into the Napo
River, where Moises had grown up. The effects on the river animals
were undocumented. "The only instances of accidental dolphin mortal-
ity that we are aware of," Vera and her husband, Robin, had written in
their landmark paper "Biology, Status, and Conservation of Inia," "are
associated with oil exploration." João Pena had told them that Ameri-

can companies contracted by the Brazilian government agency PETRO-RAS were exploding large charges of several hundred pounds of dynamite along the length of the main tributaries of the Amazon. The explosions were followed by clouds of fish, turtles, and dolphins. They floated up dead, like ghosts rising from the Encante.

But tourists are seldom aware of these events, and they still flock to the Amazon as a Mecca of wildness and wonder. And visitors like ourselves, too, can unwittingly spoil the very wildness they seek. As we approached Roxanne's camp, suddenly it seemed that motorboats filled with blond gringos were everywhere, zipping along fast enough to overturn a villager's canoe. There are six lodges situated along this five-mile stretch of river, a concentration that has served to increase the population of local people attracted by the wealth tourists sporadically provide. The village of Pôrto Miguel, located at the center of the string of lodges, has grown from a population of about 150 ten years ago, to over 1,000 today. With the increase of population has come an increase in hunting and farming, and the tallest trees along the river are now the weedy cecropias, evidence of secondary growth after older trees have been flooded or cut down.

At Dolphin Corners, we were met by Roxanne's camp manager, David Olive. Slender, bespectacled, graying, and obviously weary from his duties, he was nonetheless delighted to have visitors who did not need medical attention. His young wife, a local girl, was away visiting her mother, leaving him in charge of the camp and its human and animal residents. Ramon, a deaf-mute with crippled hands and feet, helped David care for orphaned animals; Wilder helped with boat motors; and Samuel was the camp's chief carpenter. In the camp's round, thatched dining hall lived a young orphaned toucanet named Bottomless, and two canary-winged parakeets, also orphans. More birds lived outside, some caged, others loose. Penelope the guan stalked about on elegant, storklike legs. She liked to perch on the rails along the walkways, from which height she looked down upon people with an imperial red stare. On a harness hooked to a dog run over the grassy yard, a woolly monkey named Carlotta immediately wrapped me in her four limbs and long, prehensile tail. Her body felt feverishly warm—her normal body temperature, David later told me, is 102 to 103. I loved her embrace, but eventually had to pry her loose.

When we returned to the dining hall, out came two adorable baby uakari monkeys. With naked, bright red faces, auburn coats, and funny stump tails, they resembled tiny orangutans—although locals, having never seen orangutans, say they resemble sunburned tourists. They were named Amadeus and Ludwig. They touched everything, first with their lips, then their tongues.

As the babies hurled themselves around the room, David told us his story. He came from the States to Peru thirty-seven years ago, to manage the Holiday Inn in Iquitos. He met Roxanne there in 1980. She walked up to the reception desk with an ocelot in her arms. It was an orphan she had rescued from a restaurant, where it had been kept chained as a sideshow. "I told her she could have it, but it would have to stay in the bathroom," he recalls. The next time they met, the Holiday Inn had folded and David was working at a lodge called Amazon Village. David had trained in the Army as a paramedic, and Roxanne asked him to manage her newly purchased camp and run its free clinic. The pay was bad, the equipment meager, but he took the job. "In Korea, we did some nasty things. In Nam we did some nasty things," he explained. "Now I want to give something back."

The clinic now treats patients from fourteen area villages; last year he saw 1,260 people for ailments as diverse as diabetes, conjunctivitis, complicated childbirth, broken bones, rashes, malaria, worms, ten cases of snakebite. Locals can't believe that the treatment is free. "What do you charge?" they always ask him. "Today I'm charging double," he sometimes replies. "Give me a big smile." Gratefully, they try to repay him anyway; one child brought him a bandanna filled with eggs.

His supplies, compared to his patient load, are still deficient: he has no anesthetics whatsoever. His snakebite antivenin expired in July 1995. He was down to nine bottles of antibiotics, six enemas, a single jug of milk of magnesia. He gratefully accepted our supplies, which Dianne had procured from a neighbor who works at a hospital.

David entertained us with his stories during an enjoyable dinner. He told us about some of the dolphins who frequent the camp: Sadie, who, with her baby, will approach him within six feet; Isabel, whom he described as "like my wife—zipping around all over the place, back and forth, back and forth"; Marie and Daphne, who look identical except Daphne has a white scar on her dorsal; Old Fart, the one who spouts the loudest and sounds like the Jolly Green Giant after eating all his own beans. He told us again the story of how dolphins saved Roxanne from the bull shark, which he had heard but had not witnessed. And he told us a story Roxanne hadn't: how one day, a male dolphin had rammed her while she was swimming, knocking her hard with his snout under her left breast. It may have been the breeding season, and he may have mistaken her for another male; or, I thought, had the dolphin been using his sonar, he would have seen that this was the precise location of her heart. "They are very strong animals," David said. "We shouldn't underestimate them."

Roxanne, David explained, spends most of her time in the States, raising funds—which was why we were not able to hook up with her in Peru. Meanwhile, David stays here, preoccupied with day-to-day disasters: An anaconda killed all his chickens. Pit vipers hide in the bamboo. He keeps cutting it down, and Roxanne, who thinks it is beautiful, keeps planting more. His latest difficulty, he said, was a plague of vampire bats in camp. That was why the green mealy parrot in the cage outside is so lonely, he said; vampire bats killed its companion. ("This was not in the brochure," I whispered to Dianne.)

At that moment, a cockroach jumped into my face and then ran down my arm. I flicked it away, directly onto Dianne's plate, from whence she then flicked it across the room, like some kind of insect badminton. And then a mammal ran over my foot.

"A rat," David said.

I discreetly folded my legs up on the chair, Indian-style.

"One other thing," David said, as we were about to go off to bed. "There might be tarantulas." One had dropped onto his chessboard in the middle of a game the other day.

"Oh my God," said Dianne, "the one thing I am terrified of."

"I said, there *might* be," David countered.

"In our *room*?" asked Dianne.

"Well, tuck your mosquito nets in tight."

We were unmolested by tarantulas that night. Instead, I was visited by other invertebrate interlopers: chiggers, the legacy of my cuddle with Carlotta, were now crawling beneath my skin. The itching woke me up at three that morning. But there was nothing to be done; we went out to look for dolphins. At a small, deep lake ringed by kapok trees and veiled with water lettuce and giant white lily pads whose yellow stamens were crowded with coppery beetles, we found them: We watched them till sunset. They rose and fell like waves, leaving wakes of quicksilver that lingered on the lake the way dreams linger in the soul in early morning. The lake rang with the weird cries of horned screamers, corpulent black and white birds whose call sounds like a cross between a monkey and a goose, a song alternately gulped and honked and alarmingly loud.

By evening, my chigger bites were already infected. Each bite sported a hot, red aureole big as a quarter, with a swollen circle of pus in the middle. It looked as if my entire midriff, front and back, was covered with nipples.

Everyone at camp examined the bites with great interest and offered different cures and prognoses. David said to cover them with sheep dip or nail polish and they'd be gone in three days. ("Alas," I told him, "I left my sheep dip at home!") Unfortunately, Dianne did not have nail polish—normally she would have, to touch up her toenails, but she'd left it home. Dianne felt DEET would work. Moises said to rub the bites with lemon and they would be gone in twenty-four hours. So after a full day of excruciating itching, now was the moment I had been waiting for: with a plate of cut lemons, a bottle of DEET, and Dianne to do the bites on my back, we headed off to the cold, dark shower to do battle with the chiggers.

Just at the threshold of the shower, I noticed a two-foot-long dark red snake on the raised board walkway. I stepped over it. I hoped Dianne, who was just behind me, wouldn't notice it.

"Fuck!"

But she did.

"Sy, there's a giant goddamn fucking *snake* on the walkway!"

"I was hoping you wouldn't notice," I said.

"Jesus, it looks poisonous! It's a giant, poisonous, goddamn fucking snake!"

"It *might* be poisonous," I conceded. "But it's not even in the strike position. Oh, come on—I'm crawling with chiggers here! Just step over it! I can't do my back without you!"

Dianne, however, refused to step over the poisonous snake. Instead, she went to get Moises. The snake must have sensed something was amiss, for it wisely crawled under the walkway and dropped into the water before Moises arrived with his machete, eager to kill it. A snake that color, he said, could only be an aguaje machaca, a venomous water viper.

Finally, Dianne stepped over the threshold to the bathroom. In the dark shower, with mosquitoes swarming from the drain and the ceiling corners littered with the shed exoskeletons of large spiders, I stood naked and shivering as we dosed my angry skin with insecticide and fruit juice and cold river water. "You realize, don't you," I said to Dianne, "that I have been looking forward to this moment all day."

The trip home was exhausting and itchy. We were both nervous about taking Ayahuasca. So, it seemed, was everyone else in camp. Steve had gallantly promised to attend the ceremony. I was glad, because he knew how to restrain crazy people. I had no idea how Dianne or I would react to the drug. I remembered all the stories I had heard in high school and college about people on acid who jumped out the windows of tall buildings.

Moises would stay with us during the ceremony, too, and also Graciella and Gladys. The women, ever practical, wisely brought an enormous plastic basin into the room, and explained this was where we were supposed to throw up or excrete any other material the Ayahuasca brought forth from our bodies. We were warned not to leave the room, or to expose ourselves to any light, or it could permanently damage our eyes. Nor should we speak. But Ricardo would know if we were in trouble, Moises assured us. "If you be scared, if you throw, Ricardo will blow smoke on your head to help," Moises promised. "The Mother of the Vine protect you. She will watch you for three days. And then you are free.

"You ready?"

Ricardo began the ceremony. Blowing incense over the bowl, he handed us the Ayahuasca.

It was warm and thick as blood, and bitter as bile. We both drank three deep draughts. Each swallow brought a shudder, as a tidal wave of nausea swept us both. I had planned to wait for its effect repeating some kind of calming mantra: "Mother of the Vine, come to me," or something like that. Instead, my mind chanted desperately, "Don't puke, don't puke, don't puke, don't puke. . . ."

Ricardo snuffed the light and began his whistling chant. I held Dianne's hand with my left and Steve's with my right, feeling my body swirl with the drug. I waited for the visions to take me.

The first vision came quickly. Inside of my closed eyelids, it materialized: a jaguar in the moonlight.

I realized this was a powerful omen. The Barasana people of Colombia consider jaguars the intermediaries between earth and sky, life and death, spirit and soul, water and land—and no wonder: This largest American cat is master of the worlds, for it can swim powerfully, climb trees like a leopard, and is active day and night. The Kayapo-Gorotirés say the jaguar originally owned fire; now you see it only reflected in its eyes. In fact, the Kayapos called the first flashlights they ever saw "jaguar eyes." Among the Tirios, with whom my friend Mark Plotkin has lived, a shaman may take on the form of a jaguar, in whose form he is capable of unbridled ferocity; the Shipibo say a man who wants to become a powerful shaman will see a vision of a huge jaguar enter his anus and emerge from his mouth. Thereafter, spells cast by the shaman are so strong they cannot be broken.

It seemed I watched the jaguar for many minutes. Its silver, devouring eyes stared directly into mine. It didn't move. I didn't move. And although I had seen this look on the faces of tigers, and knew what it meant, I was completely unafraid, for I was aware that this was a spirit-jaguar. If it chose to devour me, I would submit.

Had the jaguar appeared to tell me something? No words came into my head, no message. Its presence was enough. And then its eyes dissolved into my own. I saw my own eyes staring back at me under my eyelids.

This eerie vision, after many minutes, dissolved, too, as I felt Dianne let go my hand. She was going to throw up—something she was desperately hoping not to do. Since vomiting is something not even she can pull off with style, she was intent on bypassing the basin, retching privately—outside, over the rail. But when she made her escape out the door, the two other women followed to make sure she didn't fall into the water. I tried to get up, but noticed my legs were no longer attached to my body.

Over the next three hours, my main preoccupation was trying not to throw up. Ricardo must have known this, for at the times the nausea was worst, he blew incense over my head, and the rhythm of the shaka palm quickened. Steve and Dianne later told me they felt me shaking with the effort to keep the drug in my body. I wanted the dolphins to come. I wanted to see the Encante. I waited for some message, some vision.

Finally, I could stand it no longer. I crawled drunkenly to the door to retch over the rail of the walkway. I didn't feel any better. By now the drug was in my blood, not my stomach, and so what was making me sick was in my brain, I reasoned hopefully; perhaps there would be more visions. Supported by the contingent of staff who had raced to help me, I returned to the little room.

The ceremony went on and on, and so did the nausea. I began to think that of all ailments I had suffered, this was the worst. This was worse than the time I got seasick shark-tagging with a biologist: Just as I was turning green from the chop, he had decided this was the time to lure the sharks, and poured overboard a bucket of fish blood. This was worse than all the diarrhea I'd suffered in four trips to India. This was worse than the time I got dengue fever. I'd been with Dianne at a Dayak tewa in Borneo, drinking rice wine flavored with the corpse of a fetal deer out of a human skull—but at least with dengue, I had gone unconscious for three days, and missed the discomfort. Dianne had watched over me then, and before I passed out, I gave her instructions that if my body began to stink she should call my husband to tell him I'd died.

Finally, Ricardo stopped chanting, and Moises pronounced it was over. In the morning, Ricardo would interpret our visions and dreams.

"How do you feel?" Moises asked us.

"Fresh as a daisy!" I answered. "Fan-tastic!" chirped Dianne.

"Did I mention I feel like I am going to die?" I said to Dianne and Steve when everyone else had left the room. "Maybe you are," Dianne replied. "That would be an improvement," I said, and we three dissolved into a fit of giggles. We felt so miserable it was funny.

Hours later, still floating on great waves of nausea on my bed in our room, I received my final vision. "Dianne—do you see anything?" I asked.

She had been dreaming sporadically, she said: pleasant images of canoes sliding over taffy-colored water, the dorsal fins of pink dolphins.

"Well, you won't believe what just showed up under my mosquito net," I said.

"What?"

There, hovering above me, as clear as a broadcast of *Star Trek,* was the perfect replica of the Starship *Enterprise.*

A FORTRESS OF ORCHARDS

"So, what was the *significance* of all this?"

Three weeks later, as I was telling the story of our Ayahuasca experience, my dinner companions were skeptical. They included three biologists and a lawyer, and all were high-powered professionals from Chicago. My friend Paul Beaver, who has a Ph.D. in biology, called this group "the Scientists" and considered them big guns. He thought it important for me to get to know them. They were all members of the board of directors of the Chicago-based Rainforest Conservation Fund, which has supported the 800,000-acre Tamshiyacu-Tahuayo Community Reserve just outside Paul's camp since its establishment in 1991. It was the first reserve in Peru to be run for the benefit of the people living on its periphery as well as its plants and animals—an extremely unusual approach, I realized, but one which, possibly, held out realistic hope for preserving the pink dolphins' world.

Except for its current president, none of the board members had ever laid eyes on the reserve or met any of the people their funds were meant to support—until now. The group had decided to spend two weeks that June exploring the reserve they had helped to found. Intrigued by the group's premise, and eager to learn more about the rain forest from the scientists, I gratefully accepted the invitation to accompany them.

While Dianne was on an errand in England, I now found myself back in Peru, sitting with my new friends at one end of one of the long wooden tables at Paul's lodge, drinking beer and Cokes in the flickering lamplight and telling each other our stories.

Jon Green, a tropical biologist with a specialty in invertebrate physiology and a professor at Chicago's Roosevelt University, thought taking Ayahuasca was dangerously foolhardy. Jon was not a man of excessive caution: With his wife, Joy Schochet, a developmental biologist, he had spent most of two decades in Thailand and Malaysia, elbowing aside crocodiles and venomous snakes to sieve worms and crabs out of the mud. His question about the significance of my experience was tongue-in-cheek: To his mind, the only significance of the visions was the drug's physiological disruption of normal brain chemistry. But Jon, like me, was a *Star Trek* fan, and he wondered what the shaman had made of my vision of the Starship *Enterprise*.

Ricardo had interpreted the jaguar vision first, I explained. The jaguar, he had told me, had been sent by the dolphin spirits as my defense on the journey. "Everywhere, in your house, in the canoe, walking on the ground, you will have the jaguar as your defense, watching over you like a shaman," he had told me through Moises. "The jaguar is very powerful. Every day in the Amazon, a thousand thousand people going to die—but you have your defense. Maybe the Mother of the Vine knows about the tigers"—and here he was referring to my work in India—"and that is why she gives the jaguar."

My own eyes staring back at me beneath my eyelids Ricardo had ignored. (I later learned from a friend back in the States, Cindy Thomashow, who had studied shamanic trances, that an image like this is a common element in vision quests, meant to focus introspection before journeying further.) But the Starship *Enterprise* was another matter. A long train of Spanish and Quechua had followed my account of the vi-

Tamshiyacu-Tahuayo Community Reserve

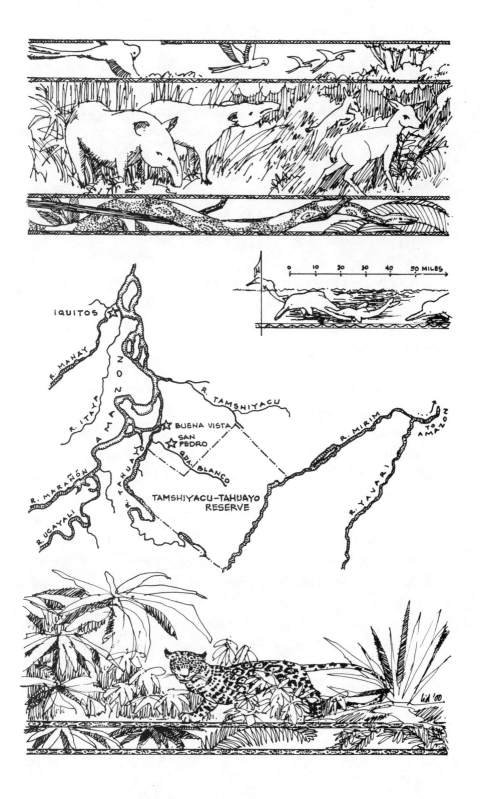

IQUITOS

R. MANAY

R. ITAYA

A M A Z O N

R. TAMSHIYACU

BUENA VISTA

SAN PEDRO

QDA. BLANCO

R. MARAÑÓN

R. TAHUAYO

TAMSHIYACU-TAHUAYO
RESERVE

R. UCAYALI

R. MIRIM

R. YAVARI

TO AMAZON

0   10   20   30   40   50 MILES

N

sion, as Moises tried to explain to Ricardo what a spaceship is. Once he understood, though, its significance was clear. Normally, the spirits send a canoe in which to travel to their realm, he explained; but because I was a foreigner from a country so technological, they had instead sent a spaceship. Or perhaps, I later thought, they had sent a spaceship because I still had so far to go.

Dave Meyer, one of the top medical malpractice lawyers in Chicago and a skilled naturalist, listened to my story and laughed uncomfortably. "But you don't believe any of this, do you?"

Did I?

"I know Don Ricardo was not lying to me," I said carefully. "I think that what he is saying, and what the people believe about the dolphins, and the way they see the world, is true—but maybe not in the literal sense that we understand."

"But this nonsense about the Encante, and dolphins turning into people," Jon broke in. "These dolphins are a biological species—they don't have magical powers!"

"Don Jorge told me he has seen the dolphins come out of the water," said Greg Neise. The Fund's charismatic young president, Greg was the only member of the board who had been here before. Greg had often stayed and traveled with Jorge Soplin, an elderly local expert on uses of the native plants. "He says they come out and stand on their tails and have the most enticing female figures you've ever seen. At the Quebrada Blanco, one did it to him, but he turned around and split—he knew it was really a dolphin."

"Oh, come on," said Dave.

"Well," said Greg, "strange things happen here."

Greg was the only one among us without a college degree; he has learned what he knows by experience. A big, bearded man in his midthirties, he ties his long blond hair in a ponytail, and his third finger bears a tattooed cross, a souvenir of a brief association with a Chicagoland gang. Greg taught himself photography and computer design. Income from the latter enables him to travel widely in the tropics, particularly in Costa Rica and Peru.

"This happened to me," Greg continued. "We were walking along the Quebrada Blanco, me and Don Jorge and some others. And we understood what the other was thinking. Without speaking, he directed me through the most difficult landscapes. When one of us saw a group of animals, the other knew it. We had some kind of awareness that we in the city do not have. There was some form of telepathy. And it goes against everything I think and believe. And it happened among people who—why, I didn't speak a lick of Spanish then and half the words were in Quechua anyway. Yet I knew what they were thinking as if we were talking. And there were no hand signals."

Gary Galbreath, a paleontologist who teaches biology at Northwestern University, abhors superstition, and subscribes to the *Skeptical Inquirer,* a magazine that debunks pseudoscience like ESP. But he is also a diplomat, RCF's president before Greg. He looked at Greg's story, as he does most things, from an evolutionary perspective. Perhaps Greg had been reading body language, he offered; as cooperative hunters, our ancestors silently communicated with one another for eons. "I bet we can still respond on a level we don't normally use," Gary said.

Greg continued: "I didn't even realize this was happening until we got back to Iquitos, and it was suddenly gone. Jim, too, has had this happen with people he didn't know." Jim Penn, the field director for the reserve's agroforestry project, has worked here in the reserve for more than a decade, sometimes living off money he makes painting houses in Chicago in the summer. His wife, Doris Catashunga, is Peruvian, and the people who live in the villages that ring the reserve are among his closest friends. Jim, with intense brown eyes the color of the earth and skin that seems perpetually sunburned from his hard work in the field, had joined our small group at the end of the long table. So far, he had listened quietly, but now he spoke: "The thing you've got to remember, is these people are as smart as you and I. There are reasons for their beliefs. And if you dismiss what they say, they'll pick up on it right away."

In the morning, we set off for the reserve. In two motorized aluminum canoes, we cruised up the Tahuayo to the seasonal white-water stream called the Quebrada Blanco. Our group numbered fourteen: Besides the RCF board members, Dave had brought his teenage son and nephew; Jon had brought two of his students; and a handful of other RCF members had come along, including wildlife photographer Jim Rowan.

As the waters narrowed, the forest seemed to quicken: Partially submerged trees trailed curtains of vines. The air, hot as breath, shuddered with the calls of parakeets, the glittering wings of dragonflies. Strangler figs flowed over their hosts like melting candles; ferns uncoiled, curling and twisting like dancers. Trees hung heavy with the nests of ants, wasps, termites, birds.

"There's never been a botanist in here," Greg yelled to me over the noise of the motor. "Who knows how many new species we're looking at!"

Actually, the leading expert on Latin American botany, Al Gentry, had come here, shortly before he was killed in a 1993 plane crash in Ecuador. He'd told Greg the place was very odd. He'd seen plants here that were not known in Peru. Ornithologist Ted Parker had the same reaction: "These birds shouldn't be here," he'd told Greg. Some of the species he saw were known only from Brazil. More than seven hundred species of birds have been recorded in the reserve—so far.

Greg and I had talked about this over the phone in the States, before we had met. "It's a really strange area," he'd told me. "Rivers flow in opposite directions, even though the land is flat as a pancake. I've seen the water in the river go up and down thirty-six feet in twenty-four hours. Tree ferns are growing in the understory, which you don't find in lowland rain forest. I've never seen anything like it. It's like something from the Cenozoic."

Unlike most of the Amazon, the Tamshiyacu-Tahuayo area has been a rain forest since before the river itself was born. Based on fossil evidence, paleogeographers believe that during the last 50,000 years, the Amazon basin became cooler and drier at least four times, and most of its rain forests changed to savannas. Fewer than a dozen small patches in the entire Amazon basin—comprising perhaps 15 percent of its size today—remained wet and warm enough to preserve the ancient lineages that thrived in the lush, steamy world of the Eocene, the dawn epoch of the Cenozoic. This was one of them.

"Where is the reserve? Are we in the reserve?" I kept yelling over the motor. Greg laughed. "Someone else who came here said, 'I went to your reserve and I didn't see any conservation work going on—it was just pure forest.' And I said, 'Exactly!' "

There are larger reserves in Peru—Pacaya-Samiria, on the western side of Loreto province, covers 5 million acres, six times the size of Tamshiyacu-Tahuayo—but none is more pristine. Because of its beauty and diversity (including a population of more than 700 pink dolphins, studied by the chair of the Cetacean Specialist Group of the World Conservation Union, Stephen Leatherwood, until his death from cancer in 1997) Pacaya-Samiria has been designated as a protected area since the 1940s. But in Peru, as in much of the Third World, merely establishing a reserve does not guarantee its protection; regulations do not ensure that the government enacting them will either enforce or even obey its own laws.

Much of the land is protected on paper only: During the 1970s, oil companies drilled no fewer than seven wells inside its borders, and one of them is still pumping. In 1991, only an intense campaign by American environmentalists, including the Nature Conservancy, kept the Peruvian government from granting Texas Crude a license to further explore and extract oil from the reserve. The Peruvian Fisheries Ministry has, in fact, granted commercial fishing companies licenses to seine the rivers inside the reserve with their giant gill nets, drowning pink river dolphins, tucuxis, manatees, and giant river otters. Despite the law and twenty park guards to enforce it, some 10,000 people, many of them with cattle, are living inside that reserve, besides the 60,000 in villages surrounding it. And the waters are silting with runoff from the 180,000 acres of rain forest that were illegally cut to the west in the Huallaga River basin to grow coca for cocaine.

But the Tamshiyacu-Tahuayo reserve has no such scars. There is no commercial logging; no oil drilling; no cattle grazing inside its borders. No sign announces the boundaries of the reserve, either. No forest guard stood watch as we entered—only a pygmy marmoset, clinging with orange hands to a spine-covered chambira palm. Its black-ringed, tawny tail hung down like a lizard's.

Moises and Mario pulled the canoe near shore and we waded through the mud. A narrow path, cut by Mario, led into a twilight world: though the sun had been bright on the river, here in the forest the light was a soft, dim green. Crickets sang as though it were evening, or right before dawn.

We walked slowly, not in apprehension but amazement. Everywhere we encountered creatures none of the biologists had seen before: climbing ferns spiked with four-inch thorns; a low-growing herb with ginkgolike leaves, fringed with brown spores—which Moises told us, incredibly, was a palm. Jon and Joy stood transfixed by a delicate white fungus growing on a fallen machimango fruit. "George Lucas wouldn't have had to use his imagination," said Jon. "He could have just come here. Imagine this thing the size of Jabba the Hutt!"

With Moises' and Mario's help, Mark, Greg, Jim, Dave, and the boys dove after frogs, spiders, millipedes, and beetles. At the day's end, they had collected eleven Ziploc bags of animals to take back to camp to be identified and photographed before they would be freed the following morning. Among them was a brilliant, two-inch green and black frog unlisted in any field guide. But Moises knew it, though not by any Latin name. It was one of the frogs the people on the Napo River use to produce poison for arrows and blowgun darts. In poison glands scattered all over the body, frogs of this family, Dendrobatidae, produce

The opera house in the jungle, the Teatro Amazonas.

In the wet season, the streets of the towns become waterways.

Flooded forest by day.

With a bulbous head and a tube-shaped beak, the pink dolphin seems poised on the brink of becoming something else.

The Amazon at sunset.

Chuckles, the only pink dolphin in North America, at his exhibit at the Pittsburgh Zoo.

The most flexible of all whales, the pink dolphin can touch snout to tail.

Above, right: A red-bellied piranha, whom we later ate (rather than the other way around). (Photograph by Sy Montgomery)

The pink dolphin's eyes are small but fully functional, and will look you directly in the face.

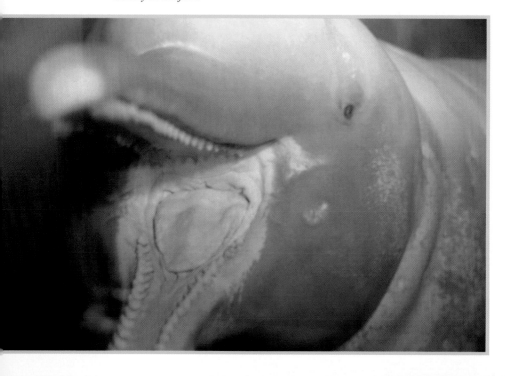

Left: Abundance of the Amazon: a fruiting cashew.

Right: Fungus glows like flame in the rain forest's filtered sunlight.

In the wet season, the nests of bamboo rats are at eye level when you stand in your canoe.

Below, left: A rare baby tamarin awaits illegal sale in Iquitos.

Below, right: A red uakari monkey.

Above, left: A spider clutching her egg sac hurries over a bird's nest.

Above, right: Epiphytes like this bromeliad perch harmlessly in treetops.

A man with a woolly monkey at the Iquitos market.

Near tragedy: the rescued egg of a white-banded antbird in a mimosa tree.

A hoatzin. Young hoatzins sport claws on their wings.

some of the most potent toxins in the world—so poisonous that one species, the bright yellow *Phyllobates terribilis,* will burn your skin if you touch it. One frog can provide enough secretion to poison fifty darts, which remain potent for a year. The poisonous secretions of another frog, which the Mayorunas call dav-kiet!, is used to produce *sapo,* a paste which is applied to the skin through a burn wound. Though at first *sapo* causes great pain—the blood rushes as if the heart will burst— its later effects are worth the ordeal. The drug heightens the senses and strengthens the body, allows you to see in the dark, sense the powers of plants, and foresee the plans of animals. The frog who gives these powers is held captive for three days before its poison is removed through gentle scraping, and when it is set free, village children run after it, wishing it a good journey home.

With the others bent over their frogs and fungi, I looked for Gary. On the long boat ride from Iquitos to the lodge, Gary had begun to tell me about some of the prehistoric animals who had inhabited South America millions of years ago. At the time pink dolphins may have entered the Amazon from the Pacific, 15 million years ago, the dominant predators on the continent were wolf-toothed marsupials called boyhyaenids, who carried their young in pouches like kangaroos, and six-foot-tall terror birds, who were essentially feathered dinosaurs, 50 million years after dinosaurs had gone extinct. Only in the previous months had paleontologists found the forelimbs of the giant birds, Gary had told me, and the discovery was a shock. Though the creatures were clearly birds, with feathers and beaks, they did not have wings: they grabbed their prey with grasping, three-fingered arms. Listening to Gary was like standing beneath some magnificent, fruiting tree, where you are showered with stories and facts and theories so exotic and delicious you swallow each greedily and wait hungrily for more. It's no wonder Gary's students love him and year after year vote him one of the best teachers at Northwestern.

But now the professor was uncharacteristically silent. He stood at the edge of a swamp. Gary's green eyes were illuminated by some inner light, focused on something in the distant past. I touched his arm. "Where are you?" I asked.

"Carboniferous," he said softly.

His words came as an echo from 300 million years back in time— back to when South America, frigid and sometimes glaciated, huddled with Africa, Australia, India, and Antarctica in one giant continent near the South Pole. But the landmass that would become Europe and North America, hovering over the equator, contained vast, dreaming swamps: the world's first rain forests. Back before the plants had learned to flower, back before the animals learned to lay hard-shelled eggs, back before hearts beat with four chambers, day-long twilight glowed green through the scalelike leaves of thirty-foot-tall club mosses. Two hundred fifty million years before the first grasses evolved, tube-shaped plants called horsetails grew skeletons of silica, and ferns danced and curled into fifty-foot trees. Six-foot, flat-headed amphibians lounged in blood-warm shallows. The ancestors of the millipedes now in our Ziploc bags grew six feet long back then, and dragonflies soared on the wing spans of macaws. The Carboniferous was the cradle of Eden. During its 75-million-year reign arose the three lineages that would lead to the vertebrates that dominate the land today: the salamanders and frogs, the reptiles and birds, and the four-legged, lizard-like animals called synapsids that, many millions of years later, would lead to dolphins and to scientists.

Dawn itself, as Gary saw it, was preserved in the green glow of this rain forest swamp. This place preserved not only the strange, ancient pink dolphins in its rivers, but the future glimmer of dolphins encoded in the genetic material of the first fleshy-finned fish ancestors who emerged from shallow swamps like these. The ferns and millipedes had outlasted the driftings and collisions of continents; the dragonflies had

Time traveler: evolutionary biologist Gary Galbreath.

(Photograph by Sy Montgomery)

survived desiccation, asteroids, cataclysmic extinctions. But could this place now survive the apocalypse of human greed? In this era of bull- dozers and oil wells, gill nets and cattle ranches, what could preserve its future?

This was the task that the Rainforest Conservation Fund took on when Jim Penn first approached the group with the idea of protecting a reserve. Previously, the group had contributed money to projects in Costa Rica and New Guinea. "We were looking for a project that wouldn't happen if we didn't do it," said Gary, who had been president when Jim came to them with his idea. RCF could not afford to finance

guards, or to open and staff an office in Iquitos. But this, Jim pointed out, was fortunate: "What you think conservation is in the United States doesn't work here," he told them. In fact, almost everything we think of as effective conservation in the States has backfired horrendously in Peru. Jim knew from his years of work here that laws don't protect the land. Officials are commonly bribed. Guards do not protect the land. In fact, park guards are often the most notorious poachers. Ecotourism seldom protects land. Unlike at Paul's lodge, where the staff is employed full-time and have health and retirement plans, at most lodges, the support staff are paid only during the busy season. To supplement their sporadic income, they work part-time for logging operations and poach game to eat and to sell at the market in Iquitos.

To the local people at Tamshiyacu-Tahuayo, Jim explained, "conservation means gringos, fancy cars, ecotourism. Conservation is people sitting behind big mahogany desks in offices in Iquitos and Washington. Conservation is a business"—one which employs scientists and bureaucrats in the business of telling local people what they can and cannot do in their own backyards.

Instead of implementing a conventional conservation scheme, Jim asked the Rainforest Conservation Fund to secure this land's protection by supporting the people's old ways of guarding it—ways which don't sound or look like conservation at all.

Today parts of the reserve are still vulnerable; but on two sides, thanks to the people who use the reserve, its boundaries are secure. The southern half of the reserve is guarded by warring Indians, the Mayorunas and their enemies, the Remos; it is simply too dangerous for loggers or cattlemen or oil explorers to enter there. And along the western border, where 175 families live in the six villages Dianne and I had passed daily, the reserve is protected by a growing fortress of orchards.

"This is sangre de dragón," Jim says, as Don Jorge points to a sapling with heart-shaped leaves. "It means 'blood of the dragon.' It yields a red resin that heals skin burns, wounds, dysentery, and internal ulcers. It stimulates regrowth of tissue."

Jorge speaks, and Jim translates for us. Jorge is slender, wiry, and spry, with the face of an imp and a smile that gleams with dental gold. His hair is soft and curls like a cherub's just above his shoulders, with a strange white patch in the back as if he slept with his head in a shallow pool of Clorox. He does not seem seventy-one years old.

This two-foot sapling? Its name is ojé, Jim translates for us, a species of ficus, whose latex can be fermented into a medicine to kill six kinds of intestinal parasites. This spiny, climbing vine is called cat's-claw. Dutch scientists are investigating its abilities to build up the immune system of HIV patients, Jim adds. The people use the bark of the yerba santo, or saint's weed, to make a medicine to deliver miscarried fetuses.

"Everything in here is useful," Jim says. "This is exceptionally diverse, with over sixty different species."

As we follow Jim and Jorge along a slender trail, Jorge briefs us on the plants we meet as if introducing us to old friends—as Moises did in the rain forest with Dianne and me. But we are not in the rain forest. We are at the edge of the reserve. Though all these are rain forest plants, and the place looks exactly like a young forest, this is his *chacra*, planted on fallow land—an orchard at once jungle and garden.

This cedro is an important lumber tree, for making furniture. And this bush is ajo sacha, which will yield a garlic-scented tea to reduce fever and cure colds. From this cashapona, a spine-fringed, stilt-rooted palm, you can make the gating around your house. But it has another use, too, Don Jorge tells us: it's a penis enlarger. How does it work? Well, first, of course, it is necessary to scrape off the spines. Then the bark is plastered against the penis for fifteen minutes. This must be done on three consecutive nights, during a new moon. Why? Because new moons cause things to grow long and thin, while full moons cause growth that is short and wide.

Such is the case with trees as well. Jorge is careful to prune his cainito tree, which produces a delicious fruit, during the full moon, so it, too, will grow full.

Many of the trees in his *chacra* bear fruits whose very names evoke luscious motions of lips and tongue: mango, guaba, aguaje, araza. The ungurahui's purple fruit contains a seed that makes a chocolate-flavored drink; a garland of snail shells hangs from one of its branches. What are they for? I ask. "These snails have a ton of eggs," Don Jorge answers through Jim. The snail shells, he explains, inspire the tree to fruit.

I can feel the scientists silently chuckle now. But there is a loveliness in Don Jorge's gesture, a grace to it, like lighting a votive candle at church. The snail shells, in fact, may supply some calcium to the soil, an important nutrient; but we would be wrong, I think, to imagine that Jorge believes hanging snail shells works like spreading lime on a lawn. Rather, this is an offering, a sacred conversation with the plant, a living creature whom he likes and respects. For the relationship between Don Jorge and the plants here is quite different from the relationships between people and plants at home.

It is important, Don Jorge later told me, never to enter the *chacra* when you are not in the right frame of mind. No one should enter their *chacra* drunk, for instance. He handed a pepper to Dave's nephew, who casually tucked it into his pocket. Don Jorge was upset. No, said Jorge, you mustn't carry it in that way. It will bring bad luck. You must carry it in your hands, with respect.

Paul's business partner, Suzy Faggard, who accompanied our group to the lodge, had made a wise observation about the people here when we'd spoken earlier. No wonder people here feel close to nature, she'd said, because in the rain forest, they are literally enclosed by it. In its twilight green glow, beneath a canopy of leaves that obscures the sky, one does not imagine the blue eye of God staring down at them from heaven; instead they live encircled by a leafy world, as if by a mother's arms. Although, after two hundred years of evangelism, almost everyone here is nominally Catholic, their traditions are older, and closer to the earth. Miracles do not come from the sky, but they come commonly from the water: magic springs forth in the form of the dolphin, the anaconda, the whirlpool. And power dwells in the plants. For communion in church, the people have learned to drink the blood of Christ; but in a more ancient communion, they drink the blood of plants, to transport them back to the real world, the world that isn't a dream, the world in which we can still hear the voices of animals and spirits speaking to us, and where spirits send canoes and spaceships to carry us on our travels.

In other cultures, spiritual ecstasy is induced by fasting, or song, dance, or ordeal. Here, ecstasy's source is green. The vine snaking up this tree, Jorge tells us, is ayahuasca. He keeps it, he says, in order to see visions and dream dreams. To help it grow, he places in its boughs offerings of his wife's hair.

We follow Jorge and Jim, the professors and biologists and other professionals taking notes. Our teacher is a man who offers snail shells to trees, who was nearly seduced by a dolphin, who can speak without words. And yet, Jim describes Jorge with words one might reserve for a scientist: He has learned by experiment. "Thanks to his ability to experiment," Jim says, "he has been able to help me immensely." For more than a decade, and throughout Jim's doctoral work in agroforestry in Peru, Jorge has been one of his most important mentors. "He is not interested in money, but he's a real expert, and he loves to experiment with plants."

Over many years, Jorge has learned not only the uses of these different species, but also which ones thrive in different types of soils and under different methods of cultivation. He knows that an aguaje palm planted in the lower part of his *chacra* will grow five times as large as one planted in the upper part. Jim tested the acidity of the soil and discovered a difference in pH of only .5—a difference that can be profound. Because of his experiments, Jorge understands that guava fixes nitrogen in the soil in which it grows; he knows that spiny palms prefer the richest soils; he knows that ajo sacha is a bush in its female form and a vine in the male, and that the female form is more curative. But by offering his trees snail shells and human hair, he shows he understands something even more important: that the fates of the plants and the animals, the people and the dolphins, the land and the water, are all linked.

The plants in this *chacra*, Jim believes, are the trees that will save the forest. It's a truth well illustrated by a tree he next shows us: the palm called aguaje.

Greg had told me about the aguaje before we left the States. I had tasted its fruit with Dianne on the previous trip. It is covered with many dozens of dark red scales, and looks rather like a hand grenade. The thin flesh covering the large seed is the color of Gouda cheese. Its taste is bland, but most people do not eat it; rather, they sell it in Iquitos, where it is used to flavor ice creams and drinks. "More aguaje comes through Iquitos today than did *Hevea* latex at the height of the rubber boom," Greg had told me on the plane ride from Miami to Lima. The highest-quality fruits garner $10 to $20 per twenty-kilo sack—a price up to $2.20 per pound. So prized is the fruit that people come from all over the province to compete for the same stands of female trees in the forest.

The harvest is difficult and dangerous. The glass-smooth trunk of the palm is nearly impossible to climb and can soar for 130 feet before giving way to a starburst of spine-covered fronds and its clusters of scaly fruit. Collectors don't wait for the fruit to drop; monkeys and birds would get to it first. Instead, the first collectors to arrive cut the trees down to prevent others from doing so—often before the fruits are even ripe. Unripe fruit sells in the Iquitos market for $1.50 to $3 for a forty-kilo sack—as little as 4 percent of the price of ripe aguaje. When Jim began documenting this phenomenon in the 1980s with Richard Bod-

Aguaje at the Iquitos market.

mer (now with the University of Florida's Tropical Conservation and Development Program, where Jim is pursuing his graduate degree) and McGill University's Oliver Coomes, they called it "the race for aguaje" —a race no one wins.

"It was a lose-lose situation," Jim explained. "Every year people worked harder and harder to further extinguish a valuable resource."

For forest animals, the destruction of aguaje was a disaster. Not only do these fruits feed monkeys and birds, but they are also surprisingly crucial to hoofed animals. For his doctoral thesis, Richard Bodmer discovered that fruit, and not grasses or leaves, provides a full 87 percent of the diet of the gray brocket deer and well over half the diet of collared and white-lipped peccaries. Aguaje alone, he found, comprises nearly a third of the diet of the lowland tapir. Finding a way to protect the fruiting palms, he wrote, was a project needing "urgent attention."

Ironically, only a generation or so ago, local people grew aguaje in their *chacras*. But most people had abandoned the old way of farming. "The government-sponsored programs tell people the way they live is

backward and savage," Greg had explained. "The way to be part of modern Peru is to have cattle and corn."

Even the conservation organizations backed the cattle-and-corn schemes. If the local people could be "civilized" into farming cattle and corn, the idea went, they wouldn't need to go out into the forest, cutting the trees and killing its animals.

The idea seemed to work at first. Corn and cattle will thrive here for two or three years—at which point the aid workers move on, satisfied with their success. But within five years, it all falls apart. The crops fail, exhausting the soils in which they were never meant to grow, the cattle sicken and die, and the land, spent, may never recover. The destruction spreads like a cancer. Brazilian and American researchers, working under the leadership of American ornithologist Thomas Lovejoy, have found that small cleared patches lose up to 36 percent of their biomass soon after being isolated from surrounding forest, when large trees near fragment edges die. And to continue to farm, the family must cut and burn more virgin forest.

Under the seductive spell of the foreign scheme, the people abandoned their *chacras* at the edge of the rain forest. Once this land was no longer used, it was up for grabs. The government began to cede these lands, which no one seemed to own, to land-seekers, oil workers, and police. Outsiders burned the land and then brought in hundreds of head of cattle and set themselves up as landlords, or *patrones*. Lumbermen came from the city to cut the trees. Merchants from Iquitos flooded in to hunt game.

By the time Jim and Richard Bodmer began their studies in the early eighties, most of the local people had all but forgotten the old way of farming. But a few, like Don Jorge, remembered. And they remembered that when the aguaje did not have to compete with other trees for light, when they grew in the fallow lands of the *chacras* and in the villages, the trees often grew no higher than sixteen feet— short enough that the fruits could be easily harvested without cutting the tree down. The fruits could be plucked ripe, at the peak of their flavor and value, and the tree would continue to produce for decades more.

And so, in March 1992, with $2,500 in funds Jim solicited from the Rainforest Conservation Fund in Chicago, he began the first agroforestry program in South America to concentrate on local, native species.

"We're trying to instill a pride in methods they already know," Jim explained later. "We didn't show anybody how to plant. We worked on a way to use what they already knew."

Today, Jim is monitoring the 6,000 aguajes that thirty-three families have planted in their *chacras*. Tall and thin, his skin growing redder with each hour, Jim works side by side with the farmers, swinging a machete and a hoe under the hot sun, his body covered with mosquitoes. He could be working on forestry in the United States, cool and orderly, as he did for reforestation projects for International Paper. But no; summers he paints houses, and he works here when he can. "My hope," he says, "is that I can contribute to people here. I have family here. I want to keep working on the aguaje," he says, "because it would be immoral if I didn't." He states this as simply as if he were noting the acidity of the soil.

He is helping other families to experiment with other forest species: the tart, acidic carambola, or star fruit, as well as copuaçu, a fruit the size of an orange that tastes like a mango crossed with a guava that is now in great demand from ice-cream producers in Brazil. Fruits like guanábana, camu-camu, and sapote, as well as medicinal plants, bring premium prices at the market in Iquitos.

The trees thrive in the *chacras* because they are meant for this place. Many species, like the inga, fix nitrogen in the soil, enriching instead of depleting it. Because the *chacras* are planted with mixed species, not just rows of one kind of plant, a variety of nutrients are constantly recycled. A *chacra* of native species produces for twenty, thirty, forty years—possibly more, Jim explained. Because these *chacras* are planted on fallow land, formerly used to grow manioc and bananas, there is no need to clear wild forest. And should the family eventually abandon the *chacra,* they leave behind, instead of a charred clearing, a young, mixed forest of native trees.

But they still need the wild rain forest—and the *chacras* help them to guard it. For, contrary to many environmentalists' notions, it is not, Jim insists, the local people who threaten the rain forest, but the greed of outsiders whom locals often feel helpless to evict.

"If we don't protect our resources, who's going to do it?" Raúl Huanaquiri, the municipal agent, an elected community leader, is speaking to us in Spanish in front of the single classroom of San Pedro's mud-floored collective school. Around the school stand San Pedro's wood and thatch houses, perched on stilts at the river's edge. Once again, as at the wake for Mario's son, ours is the only canoe with a motor moored to the mud bank.

Raúl stands in front of one of the school's five green chalkboards, which hangs suspended from bamboo poles lashed together by lianas. Behind him, a map of the reserve, drawn in blue chalk, shows the waterways, the villages, the northeastern frontier of the reserve, near the border with Brazil. The scientists sit hunched in the wooden chair-and-desk sets—just as in elementary school, learning from the teacher in front of the room.

"These people know more about conservation than all the gringos combined," Jim says as he introduces Raúl. "We have a lot to learn from them—of how to live as a community, as a peaceful, sustainable community, in a peaceful way."

Jim translates as Raúl, wearing high rubber boots as protection against snakes, explains his neighbors' understanding of the reserve. "At first when the reserve was created, there was a lot of debate," he said. "Still two or three people don't like it. But everyone knows about it now. Everyone realizes we must protect our own resources. We must do it, not the government."

This the people know well. Last spring, they saw rafts of logs leaving their forest reserve, towed behind big boats from Iquitos. RCF's site manager, César Reyes, discovered that, despite the law prohibiting lumbering, Iquitos officials had formally granted nine lumber concessions inside Tamshiyacu-Tahuayo—all with official paperwork and with full knowledge of the Peruvian Institute of Natural Resources and the agriculture department. But with César serving as their professional liaison to the authorities in Iquitos, local people protested. The signatures on the official documents allowing the concessions were officially erased, the papers revoked, as if they had never existed. Only forty-eight logs were extracted before the timber concessions were formally annulled.

The people need the forest whole. They depend on it for their lives. Some thirty-five families regularly hunt in the reserve. They take the big rodents like pacas, the piglike peccaries, and occasionally monkeys and larger game like tapir. They eat the smaller animals and sell the peccaries, deer, and tapirs, as well as the pelts, in Iquitos for cash. Thanks to Richard Bodmer's studies of the populations of wild animals, they realize these resources are vulnerable—not just to outsiders, but to local pressures as well.

So now, before the professors and biologists gathered humbly in the classroom, Raúl outlines some of the changes the people of his village have recently instituted: They have prohibited the hunting of monkeys, giant armadillos, and tapirs, and restricted the take on hoofed animals like peccaries to three animals every two months per family. Locally elected hunting inspectors at two checkpoints along the river register every animal taken from the reserve; only local subsistence hunters are allowed to pass. And although the area is vast, there are no roads, only waterways, so it is difficult to pass unnoticed.

They have organized group work days, or *mingas,* to plant native trees in the *chacras* at the western border of the reserve. "The idea also," he tells us, "is to teach our kids, so in the future, they will know what a white-lipped peccary is, and also know the forest plants.

"With village protection, peccaries and large rodents are beginning to visit us again," Raúl reports, to murmurs of approval from the Americans. "We are trying to coordinate with other communities."

In fact, the reserve's methods were highlighted at a conference sponsored by the Institute of Natural Resources in Iquitos the month before. The agricultural techniques developed for the reserve were adopted for a $5 million, four-year project in Pacaya-Samiria. Jim and César would like to hire two more Peruvian agroforesters to work in the villages at the reserve's northern borders along the Yavarí-Miri River. And as Greg, Gary, and Jon discuss later, with another $50,000 they could finally afford to pay their Peruvian site manager, César, what he's worth, and give Jim, for the first time, a salary.

"The reserve has become an example," Raúl tells the board as he closes his careful, formal speech. "International people should come to see the reserve, to benefit not just the people here, but the entire country."

As the information session concludes, from his little desk in the classroom, a professor raises his hand. Jon waits for Raúl to call on him.

"I'm delighted," Jon says, "to meet such great minds in San Pedro."

TIME TRAVEL

Tell me a story.
In this century, and moment of mania,
tell me a story.
Make it a story of great distances, and starlight.
The name of the story will be Time,
but you must not pronounce its name.
Tell me a story of deep delight.
ROBERT PENN WARREN

There was a world before this one, the people's stories say. It was destroyed, by fire or flood, or most of the animals went extinct; the people turned into stones, or were devoured by monsters. Though most of the people in Tamshiyacu-Tahuayo now recite the Christian creation myth—perhaps easily assimilated because, with its world-altering apple, Genesis echoes the ancient legends of the World Tree. But they still agree with the hundreds of different Indian tribes throughout the Amazon: the world is not now as it has always been.

Once, they say, there were other worlds. Once, there were fantastic creatures, chimeras, monsters. Some of them are with us still.

In the evenings, back at Paul's lodge, Greg and Gary, Jim and Jon and

Joy and I sometimes talked of these monsters. One is called the Mapinguary, a bear-sized, hairy, stinky, ape-faced creature. Gary thought it sounded rather like a Pleistocene giant ground sloth. He related that in fact, a Boston-based scientist, David Oren, is convinced some of these animals are still living. Gary was doubtful, but Oren had mounted an expedition to search for them in the Brazilian state of Rondônia in 1994. He never saw the animal, but there he collected twenty-two pounds of what he believes to be Mapinguary dung.

Another is a creature that in Brazil is called the Water Jaguar, Onça d'água, or the Tapir Nymph, Tapir-iauara. It grows big as a cow and attacks people in boats. There are also gigantic snakes—longer and fatter than the most enormous anacondas—that dwell in the whirlpools and in the deepest portions of the lakes. Their eyes shine like flashlights. They are called Cobra Grande, and it's said that their slithering across the landscape creates the Amazon's constantly changing maze of channels. When a Cobra Grande growls, pirarucu and caimans slap the water with their tails in reply. The Cobra Grande can immobilize your boat, steal away your shadow, or take you away to the Encante beneath the water.

Some of these ancient monsters are more frightening because they can assume human form. One of them, sometimes called Curupira or Chullachaqui, is a shape-shifter who may appear as an old man, a dwarf, a deer, a frog, or as the large adult male who leads groups of white-lipped peccaries on their migrations. He is the Father of Game, and he fools hunters. His feet point backward, so that hunters think they are fleeing from him when they are actually following.

I had read about Curupira in Nigel Smith's fine book *The Enchanted Amazon Rain Forest*. A Rio Negro peasant told Smith this story of an encounter with Curupira, which he said took place fifty years ago: A hunting party had ventured up a tributary of the Rio Negro, searching for peccaries and piacava palm. They killed many peccaries—each man took more than six—and they placed the carcasses on pole platforms over the fire to roast.

That night, a stranger came to their forest camp carrying a message. "The Old Man says that those who killed five or more peccaries should give one to him," he said.

"Tell the Old Man to go hunt his own game," one hunter retorted testily.

The wife of one of the hunters, though, became alarmed. She told the group they were foolish to disobey. Her husband had killed only one peccary, but the couple gave the stranger a hind limb for the Old Man.

At midnight, thunder rumbled in the forest, at first distant, then closer. An old man wandered into the camp. With a cane, he tapped the carcasses of the roasting peccaries, and they sprang to life. Dozens of peccaries assembled in the hunting camp, as if to watch the events about to take place.

Next the Old Man went to the hammocks of the disrespectful hunters, and with his hands began to knead their bodies like dough, until inside their skins they were mushy as avocados. Then he sucked the pulp of their flesh out from the tops of their heads. He tossed their empty skins back into the hammocks. The generous couple, unmolested, heard what the Old Man muttered as he walked away: the peccaries, he said, were his children, and no hunter should take more than they need. He melted into the forest, followed by the animals he had resurrected.

When our previous companion, Jerry, had gone survival camping in the forest with Rudy, they had come across a curious stretch of parklike forest where the trees grew tall and the ground was bare of understory; it seemed as if someone had tended this land. Greg, too, had seen areas like this. They are called *supai chacras,* he said. And this was what Jerry and Rudy had been told by their barefoot guide: this land was Curupira's garden. The guide suggested they get out of there, fast. "I wanted to stay longer," Jerry had lamented when he came back to the lodge. "I would have loved to have met that guy!"

Scientists laugh at the stories of Curupira. But they have stories of their own. And they agree there *were* worlds before this one, complete with fantastic monsters. Some of them, we were to find, are with us still.

Shortly after we returned from our first foray into the reserve, I asked Gary what he had been experiencing when he had gazed spellbound before the swamp at the edge of the path.

"You could say this place takes you back in time," he told me. "The species aren't the same, but the structure of the forest here is the same. Rain forests dominated by flowering plants date to more than 100 million years ago. Rain forests are one of the oldest biomes on earth today." Temperate woodlands, he said, appeared just 33 million years ago, during the Oligocene, when the world first became seasonal; grasslands evolved only 23 million years ago, in the Miocene, followed by the first grass-eating horses; and modern coniferous forests, like today's taiga, weren't around until the Pliocene, a mere 5 million years ago. The world is not now as it always was. "If you go back to the Eocene period, at the Poles there were forests with really big leaves, broadleaf trees! And that biome doesn't exist at all anymore. Even the current glacial habitats, the North and South Poles, weren't there all that long ago," he told me. "But the rain forests were."

Gary, I found, often travels back in time. He can go there without blinking. As we journeyed through the rain forest with the other RCF board members, on foot and by motorboat and canoe, sometimes the normally loquacious professor would become suddenly quiet; he would get a dreamy look in his eyes, and one could see that he was gone. Gary sees nothing mystical about his ability, and in fact would be horrified at the suggestion that he possesses anything similar to shamanic powers. He is a scientist, first and foremost, and doesn't believe in trances or gods or powers outside those of the laws of physics.

And yet, sometimes he would say to me, "Let's go back to the middle Jurassic," or propose, "Suppose we go back into the Cretaceous," the way other people might suggest a trip to the mall. The Age of Dinosaurs is his favorite time, and in fact he will tell you this is where he spent much of his childhood, rather than in Kentucky in the 1950s. But he could see the tree-fern–shaded landscapes of *Tyrannosaurus rex* and *Triceratops* more vividly than the willow tree and the squirrels in his family's backyard. It was not until Gary was halfway through second grade—by which time he had already memorized the geologic time scale, as well as the scientific names for most of the major animals in each epoch—that an ophthalmologist diagnosed his severe myopia. With his new eyeglasses, Gary saw for the first time that the teacher was writing words on the blackboard, and that trees that had resembled big lollipops had individual leaves. But by then he was already well traveled in the lost worlds of the prehistoric past, and continued to dwell there. During our time in Tamshiyacu-Tahuayo, he would take me there with him—back into times when fish were shielded in armor, when birds flew on clawed wings; back to times of wondrous transformations, when fish began to crawl out of the water, and mammals began to crawl back in. For here in the Amazon rain forest, time is permeable; reminders of these fabulous creatures are with us still. One of them is the bufeo.

We began the day in the Devonian. Juan Salas brought us, that morning, an armored catfish, and it catapulted us 400 million years back in time, to the period immediately preceding the Carboniferous. The fish was a bit under a foot long, and covered with black, armorlike plates. Gary stared at it as if it had been plucked from Devonian seas. "Four hundred million years ago, until about three hundred seventy million years ago, one of the most important groups of fishes, the placoderms, looked much like this," he told me. The name of the group comes from the Greek *placo,* for "plate," and *derm* for "skin." "They had extensive plate armor, especially on the forepart of the body, and several of them even had little plates covering part of the eye sockets. There was a huge armored fish called *Dunkleosteus.* His skull is several feet long! This armored situation was much more common then. The placoderms were very badly hit by a mass extinction near the end of the Devonian period. Without that large-scale extinction, they might have continued to be a really important group of fish."

Today's armored catfish belong to the family Doradidae, which evolved their armor separately from the placoderms. One of them, called the rock-bacu, is so named because its hard covering gives its flesh the weight of rock. Some of these fishes' thick, bony plates—their plate-skins—support backward-curving spines, each serrated like a saw, which they can erect like a trident. Later, Dave Meyer, the lawyer, discovered this to his sorrow, when he tried to hold one of these animals, pinning it in shallow water with the palm of his hand. He spent a quarter hour trying to dig the spines out of his flesh.

The placoderms had to contend with predators more fearsome than lawyers. Among the giants in Devonian seas, Gary told me, were huge, eel-like sharks with bizarre jaws: the upper jaw was like a Frisbee covered with teeth. They rammed their prey, severing the body in half. Back then, the great monsters all lived in the seas. The earth's scattered

lands were peaceable kingdoms of giant club mosses growing thirty feet tall, towering, arrow-straight horsetails, and seed-bearing plants that bore fruit but no flowers—the ancestors of the conifers. Until the twilight of the period, the landscape of the Devonian was untouched by the fleshy fins that would later become toes and feet. But in the genes of these ancient fishes were the unlikely future of the land animals: the amphibians and reptiles, the hoofed animals and the terror birds—and, surprisingly, the mammals who would much later venture back into the water as whales and dolphins.

Gary and I ate breakfast 360 million years later. Over eggs and beans, we had planned our travels to the Jurassic. But first we would pass through the Carboniferous in our canoe.

"Three hundred million years ago, a good part of the world was like this. Most of North America and Europe was near the equator, and there were these huge swamps. . . ." The flooded forest is studded with snags capped with bromeliads. "But there were no flowering plants then," Gary says to me. "The trees had scaly bark and little scaly leaves. There were no bees or beetles or butterflies, but plenty of dragonflies and roaches. And there are fish, lots of fish, including the fish that our ancestors had come from. . . . It is a lush, steamy, swampy time. . . . I've thought about this ever since I was a kid." Gary has switched to speaking in present tense, for this is where Gary is now—back in the Carboniferous. He does not notice the spines protruding from the carachama palms as our canoe scrapes past them; there were no palms in the Carboniferous. I have to pull his hands away from the gunwales of the canoe so his fingers aren't impaled.

Gary tells me what he sees, but I cannot: The air is thick with dragonflies flying on wings big as seagulls', and huge stoneflies, the ground stalked by springtails, cockroaches, scorpions, centipedes, millipedes. The first amphibians have appeared, big as men. Forests of lycopsids,

The armored catfish, a prehistoric look-alike.

with their scaly leaves, and trees of horsetail and fern rise and then drown in cycles of inundation by shallow seas. The Carboniferous is known as the Coal Age because the flooding knocked over the trees and formed coal deposits over 3,000 feet deep.

We are bound for the Jurassic. We are heading to Caiman Lake to see one of its most peculiar denizens: the hoatzin, a chicken-sized bird whose young possess wings with claws like a reptile's. Some scientists have linked this species with the dinosaur-era bird *Archaeopteryx,* the "Ancient Wing." The hoatzin is not a remnant of that time, but a re-minder of it. The hoatzin is actually a member of the cuckoo family, and like the armored catfishes, evolved its primitive characteristics quite separately from its prehistoric look-alike, millions of years later.

But in the genetic plans for their bodies, the fish and the birds re-member the time of armor and claws—just as our own bodies remem-ber. Like Curupira, like the botos of the Encante, we are shape-shifters: As embryos, humans are first fish. And from gilled creatures floating in the womb's warm ocean, we transform to lives more amphibian, more reptilian, until finally, we assume mammalian form, complete with tails. Weeks before birth, we are indistinguishable from fetal monkeys. We emerge as complete humans only after recapitulating our evolu-tionary history, the plan of our bodies paying homage to our ancestors.

Dusk rises from the water as our canoe approaches the spindly, leaf-less snag in which the hoatzins have built their nest. Hoatzins nest over water so that when danger threatens, the young, who are excellent swimmers, can plunge to safety; they use the two claws on each wing to climb back to the nest. Two of the chicken-sized birds hiss down at us like reptiles—as the *Archaeopteryx* would have done 150 million years ago over the still waters of Solnhofen lagoon.

Since Gary was in junior high school, he has treasured a book illus-trated by the Czech painter Zdenek Burian portraying, among its many prehistoric scenes, the artist's conception of the mud flats of that an-cient swamp whose limestones yielded the vaunted "missing link" be-tween reptiles and birds. And this is the image that floods his consciousness now: the *Archaeopteryx* glides on greenish wings in a lush, wet forest of prehistoric conifers. Three claws are visible at the wing's bend. The long jaws show tiny teeth in a pointed snout, and the tail—a long, bony tail, not mere tail feathers—trails behind it, like a lizard in flight.

In fact, the hoatzin is about the same size as the *Archaeopteryx:* about two feet long, weighing two pounds. The birds above us have long tails, plumage streaked brown above and yellowish below. Their heads bear loose orange crests, and their bright red eyes stare at us from blue faces. "Wow," Gary says softly. "This is really primeval."

We could watch those birds for centuries—but dark is coming. As our guide, George, tries to thread our canoe back toward camp, increasingly we bump into ant trees—not just cecropias, whose ants are small, but other species whose leaves I cannot make out in the dusk but pray are not tangaranga. In the gathering dark, we cannot find the ants in our canoe until they bite us. These ants bite hard. But we cannot afford to pay them too much attention as we try to duck branches and dodge spines, which are now also difficult to see in the gloom.

The water level at the lake has fallen considerably in the last few days. George finds the regular water pathways too shallow. Our canoe is heavier than usual, with four passengers: We have come with Rachel Dulin, an RCF member from Israel, and Carolynn Charleston, one of Jon's former students. This is their first trip to the tropics, and they are worried that we may never find our way out in the dark. But the worst that can happen, I tell them, is we spend the night in the canoe. Lightning pulses in the sky. OK, worst case is we spend the night in the canoe in pouring rain. Of course, we could be struck by lightning, too, Rachel helpfully volunteers.

George now seems clearly lost. Unfortunately, we have only one flashlight, whose bulb is dimming. And now we are stuck. We have hit a submerged log, and our canoe is wedged atop it. We really might have to spend the night in the canoe—uncomfortable, given the impending rain and mosquitoes, but not, as Gary points out, truly dangerous. But then I catch a pungent scent like buttered popcorn. In India, that scent was significant: it is the smell of tiger urine, which the cats spray on trees to mark their territories. There are no tigers here, but there are jaguars—like tigers, stalk-and-ambush hunters; the name comes from the Indian word *yaguar*, meaning "he who kills with one leap." And like tigers, they establish territories, which they mark with squirted urine.

I again revise my worst-case scenario, but I keep the revision to myself.

Pushing against two trees, we rock the canoe over the log. We are free. Now George recognizes a canal, and after a short passage, we emerge from the forest to the wide river, and motor home under the stars.

Later, I told Gary about the buttered-popcorn smell, and we asked Moises what creature might have produced it. It is the fluid ejected as a defense by a species of beetle, Moises said, which spews acrid spray from a flexible abdominal turret.

"A beetle!" exclaimed Gary. "Middle Jurassic!" And then he was gone again.

The present epoch brought its own delights. One morning, a four-foot electric eel was spotted swimming in the shallow water covering the camp's dry-season courtyard, which by now had dropped to two feet deep. We all leaned over the rail like passengers on a whale watch. Beneath us swam a weird and beautiful creature: scaleless and gray-brown with a red underside, fat and sleek, and capable of producing a discharge, from the electric organs in its tail, up to 650 volts. Dianne had washed her hair here a few weeks back, on a day when the showers weren't working; I wondered if the eel had been there all along, harmlessly concealed in the dark water. It may well have been. In Paul Beaver's book *Tales of the Peruvian Amazon*, he relates how a friend, wading through the muddy water at the edge of a river, suddenly froze when she felt the soft brush of a fish on the inside of her calf. She looked down and saw a six-foot electric eel swim between her legs. The eel slid by quickly, but to her horror, as she remained frozen, it was immediately followed by a second large eel, and then a third, a fourth, and fifth.

Electric eels are astonishingly common. One survey, of a white-water–flooded forest, found electric eels constituted 70 percent of the fish biomass. This one was a small specimen. They can grow to nine feet.

Every few minutes, the sinuous creature beneath us surfaced to gulp air. The mouth is so rich with blood vessels that the eel uses it as a lung; its gills are vestigial, used only to release carbon dioxide. Protecting the sensitive mouth may be one reason the electric eel shocks its prey: stunning prevents the prey fish from struggling. But often the eel doesn't bother, and instead, like a vacuum cleaner, simply sucks the prey through the mouth directly into the stomach.

Greg and photographer Jim Rowan decreed the eel must be caught and photographed. Rudy and Juan Salas descended to the pool to urge it into a net. Soon the two men were leaping about as the fearful eel tried to hold them at bay. Greg told me that up the Quebrada Blanco last year he and Don Jorge had caught an electric eel. Jorge had touched his machete to its head, enduring shock after shock in an attempt to cure his rheumatism.

"Did it work?" I asked.

"It might have." Greg laughed. "Or maybe he just felt better when it was over, because then he didn't have an electric eel shocking him anymore."

I grew increasingly eager to speak privately with Don Jorge—this man who treated his ailments with electric eels, who could speak without using his voice, and whom dolphins tried to seduce. Later, I asked Jim to arrange a visit to Jorge's house, and to translate an interview so I could talk to him about the magical powers of bufeo. To my surprise, Gary offered to come with us. A more unlikely pairing of minds I could not imagine: Gary, whose truths were fossils, literally set in stone, and

Jorge, to whom truth appeared as the wishes of plants and the va-
porous spoutings of dolphins. Would Gary, a talented teacher, feel the
need to "educate" Don Jorge? Would Don Jorge sense Gary's disbelief
and feel insulted? Neither would believe the other's stories. Gary no
more believed in Curupira than Don Jorge would believe in dinosaurs.
But both men, as I was to learn, told mirroring truths. Their stories
spoke of time and transformation, understandings that reflected each
other the way the waters of the Encante mirror the stars in the sky.

We visited with Don Jorge one afternoon at the slat-floored, stilted
house in Chino he shares with his smiling, wrinkled wife, Isabella. But
before we sat down, Don Jorge had something to show us. He led Jim,
Gary, and me to the small kitchen of the house. He lifted up the large
wooden bowl in which Isabella makes masato. Beneath it was a foot-
and-a-half-long, smooth-shelled turtle with dark eyes and a pointy
snout. Its neck was tucked to the side like a bird with its head under its
wing. Gary gasped. "It's a *Podocnemis*!" he exclaimed. "I've read about
them, but I never thought I'd see one."

Podocnemis is the name of a group of large South American freshwater turtles with necks that do not retract. Instead, these turtles tuck
the head to the side to protect it beneath the shell. They are the most
primitive turtles in the world. They may be among the few survivors
from the time of Pangaea, the great global supercontinent, whose U-
shaped landmass lay on its side with the ancient sea named after
Tethys, the Greek Titaness and sea goddess, in the middle. As continen-
tal drift began to break Pangaea apart, its bottom half became the
southern supercontinent of Gondwanaland, which contained the land-
masses that would form South America, Africa, India, Australia, and
Antarctica. Ninety million years ago, just before the Mesozoic dinosaurs
went extinct and as the flowering plants were evolving, Gondwanaland

The side-necked turtle's neck does not retract, but instead tucks to the side.

began to split, and South America became a giant island floating west. By then, however, the side-necked turtles had already appeared, accounting for their distribution today: Seven species of these turtles live in the freshwater rivers and lakes of the Amazon and Orinoco basins; an eighth species survives on the African island of Madagascar. All of them are increasingly rare.

Don Jorge recognized the turtle's rarity. He hadn't meant to catch it; it had just come up on his fishing line. What would they do with it? He and Isabella would eat it that night, he supposed. Gary and I offered to buy it. The scientists and photographers back at camp would be thrilled to see this rare creature, we explained, and after photographing it, we would let it go. Five dollars purchased its freedom. Then Don Jorge led us to the open porch, the room nearest the river, where we perched on odd-sized chairs and stools and Jim patiently translated our conversation about dolphins.

"Oh, I know everything there is to know about bufeos," Don Jorge said matter-of-factly. "When the bufeo wants to steal you away, it

transforms itself. Yes, the dolphin has the ability to stand, out of the water, and appear as anyone—even as your best friend: 'Hey, Jorge,' he'll call, 'come for a swim!' And the two of you will be swimming together, and then the dolphin is a dolphin again. And now, you are one of him. You don't even feel it, but it takes you in its canoe, and goes beneath the water." Oh yes, I thought, Gary will recognize that: you slip away, without even feeling it—to the Cretaceous, to the Jurassic, to the Encante. And in fact, Gary now listened intently, suspending his disbelief.

"The dolphins come out of the water and stand on their tails and have the most enticing female figure you've ever seen," Don Jorge continued. "And this I have seen—a woman once came out of the water in front of me. She was naked. I was attracted to her, but didn't follow. I knew it was a dolphin, and that she would take me underwater. You stay there, as an element of the water, in the city there." The Encante, he explained, is complete with cars, banks—"everything you have in Iquitos." In fact, canoeing on the river at night, sometimes he can hear the church bells toll from the Encante.

"It is a beautiful city, but you can never return," he said. "Once they take you, there's no hope. It's forever. So when I saw that dolphin, I turned around and left."

Don Jorge paused to pour beer—the custom, Jim explained, is one pours the bottle into the single glass for the other, who then drinks and then pours more beer into the same glass and passes it to another—an opportunity to share grace and hospitality when there is only one glass. "One time, at the mouth of the Quebrada Blanco and the Tahuayo, I ran into two bufeos," Jorge continued. "One was a female who lay on her back, showing me her breasts. I kept going, but then the head of a woman popped out of the water and seductively licked her lips and flicked her tongue at me. She wanted to make love with me. But the problem is, it's so fantastic, the lovemaking, that you will die of pleasure. If you're alone with that dolphin, you'll end up a dead man."

There are other dangers to encounters with bufeos, he told us. "One night recently on the Tahuayo, I was paddling in my canoe—and going nowhere! Usually, when you hit a log, you can hear it scrape," he said—and in fact we remembered that sound from our adventure on Caiman Lake. But this time, he said, he heard no such sound, and was puzzled. "What's going on?" he asked himself. "And then I knew: it must be a bufeo! Fortunately, I had a cigarette behind my ear. I smoked it, and the canoe broke loose."

"It's witchcraft!" Jorge and Isabella exclaimed in unison. The bufeo can be downright devilish. You must be careful not to offend bufeos. These dolphins also shoot invisible darts, they said, which can go into your heart and kill you. They blow the dart from the blowhole, and it can cause incredible pain. Once, Don Jorge said, a dolphin shot a dart up at him just as he was squatting to relieve himself in the latrine over the river. The dart went right up his anus! The next day, it hurt very badly. But a medicine man was able to help: he blew tobacco smoke into the orifice, and then Don Jorge was cured.

Happily, Don Jorge explained, bufeos normally seem to like him, perhaps because he plays guitar. Dolphins like music. And he knows how to repel the bufeo when necessary. You can blow smoke, or throw the tobacco directly into the water. When a woman has her period, Isabella told me through Jim, she must carry tobacco with her at all times. Otherwise, the bufeo will smell her menstrual blood and come after her. But Don Jorge is the one in more danger, she said, as he spends so much of his life on the water.

Don Jorge has seen the dolphin in many forms. "One day," he related, "I went with my parents and a few neighbors to a large lake. It was brownish water, black and white water mixed. We had decided to go fishing for paiche [pirarucu] with barbasco root." The shallow root of the barbasco vine contains a fish poison that causes the animals to float to the surface where they are easily captured. "We had two hundred kilos of barbasco root. We were well prepared to fish for two to three weeks. At midday, we were smashing roots to prepare them for the next day. We were scouting the lake when two botos appeared. And the people were saying, 'These dolphins will die when we throw the poison in.'

"We had fifty kilos of root prepared just then, at one o'clock, and two guys with nice shoes and big sombreros were walking along the side of the lake. 'Hello, how are you doing?' we greeted each other, and we told them what we were doing. Then they walked away.

"Then we went back to the lake—and the dolphins weren't there anymore. They knew that they would die if they stayed there. And the reason they knew about the barbasco was those men were the dolphins. I was ten years old then and knew what I was about. It is certain bufeos transform into people. This isn't a story. I saw this myself. It really happened.

"And another time: I was in Rio Haya at Carnival. It was February. That's a big party, a festival, and people are playing flutes. At about midnight, one of the flutists says, 'Let's eat!' A big table was laid. A young man appeared at that time, very well dressed. So the young, good-looking guy says, 'While you eat, may I try out your flutes?' And he began to play—old songs, new songs, traditional songs—and he blew them away, he was so good.

"Upon hearing him play, the band was very embarrassed, because this guy was better than them. And all the young women were totally in love with him. So the head musician says, 'Now let me play, so you can dance.' Now the people preferred listening to the young man play, but he wanted to dance. The women stuck to him like glue—but finally he had to stop dancing because the people insisted that he play again. Finally, it was three o'clock in the morning, and he said he had to stop playing to take a leak. So he went out to the docks, where the canoes were tied, and—voom!—he dove into the water!

"As this happened, the people noticed that all these dolphins were coming near: the flutist's brother and sisters and family! Then they realized that he wasn't just a good-looking guy, but an element of the water. At that time, I was twenty-two years old, and I realized, too, that he was a bufeo."

So why is it, I asked, that bufeos transform themselves into people? Are they not content to live in the beautiful Encante? Why do they try to take people away? Don Jorge thought for a moment—he hadn't considered this question before. Jim translated his conjecture: "He imagines it may be because the waters are so extensive, they need to make their populations grow," he said. "Maybe taking people beneath gives their cities more population."

Gary and I are going to look for bufeos tomorrow, I told Don Jorge through Jim. Should we take any special precautions? Should we be afraid?

"It is good to fear it," Don Jorge replies. "It is always good to fear it."

We went out with Moises the following morning to look for bufeos. Earlier, our group had visited Charro Lake; there we had seen tucuxis, but no pink dolphins. In our trips to the villages and into the reserve, we hadn't seen any, either. Unlike the month before, when Dianne and I were here, watching dolphins was not the focus of my second trip; plus, the water had dropped significantly, and the bufeos, said Moises, were following the water as it drained to the Amazon.

Moises had taken the motorized canoe out earlier to see if he could find dolphins for us. There were two, he said, a male and a female, in a deep channel leading to the Amazon. He had raced back to get us, and by the time we returned, at least one of them was still there. But all we saw was a pinkish shiver on the water's surface.

"This is what I'm trying to follow," I said to Gary sheepishly as we motored back to the lodge. "I'm sorry this is so disappointing."

But Gary understood. This was like working with fossils, he said: one glimpse, so many gaps. "There are entire organisms we know only from a single tooth."

For many years, in fact, such was the case for many species of prehistoric whales, Gary told me, the ancestors of the bufeo. Fully aquatic whales existed by the middle Eocene, 45 million years ago or so, as we know from fossil skeletons discovered in marine sediments in Texas, Egypt, and Nigeria. But how did a mammal, whose ancestors had lived for millions of years on the land, transform itself to live, like a fish, in the water? For many years, the links between the land mammal and the water creature were missing, and its millions of years of evolution were an utter blank. The origin of the whales was one of the deepest mysteries of mammalian evolution.

The answer, Gary told me, has only recently been revealed. It is a story so fantastic that, if its truth were not set in stone, no one would believe it.

"Nothing less resembles a mammal than a whale or a dolphin," wrote Giorgio Pilleri, the director of the Berne University of Brain Anatomy, in *Secrets of the Blind Dolphins,* an account of his journey to Pakistan in 1969 to collect Gangetic river dolphins there. "Nowhere is the impact of ecological factors on body shape so apparent as in whales and dolphins. Nowhere has an organism undergone such complete transformation," Pilleri wrote.

No other animals have undergone such far-reaching anatomical changes in the course of their evolution: the pelvis shrank, the hind limbs vanished, the forelimbs became flippers, tail flukes appeared without bones, the nostrils united and moved to the top of the head. The skull was reorganized completely in order to hear underwater and to withstand the pressure of deep dives. "One look at the animals is enough to show what a fantastic process it was," wrote Pilleri, "more of a revolution than an evolution."

A month after we returned from Peru, I visited Chicago's Field Museum with Gary to examine the skulls of *Inia* and *Sotalia.* Gary had once

Moises paddles our canoe through the flooded forest.

(Photograph by Sy Montgomery)

been in charge of reorganizing the museum's research collection of more than 40,000 mammal fossils; yet his reverence for bones is untarnished. With infinite gentleness, he lifted the yellowed skull of the *Sotalia* out of its long white cardboard storage box. A deep quiet settled over him, as if he were listening to the bones speak. When Gary finally spoke, his voice dropped low, an aural kneeling, like a pilgrim before a great mystery: "It doesn't look like a mammal skull," he said softly. The yellowing tag says this specimen is from Venezuela, Lake Maracaibo, 1912; the bones are still greasy from blubber. The eighteen-inch-long skull looks like a mammal skull that was made of Silly Putty, pushed and pulled all out of shape. "You look at a bear or a weasel, you're looking at the same basic parts. But here, the skull is telescoped." In front is a long, tooth-filled jaw, like a beak. The foramen magnum, the hole where the neck bones and spinal cord enter the skull, is not at the bottom, but at the back. Almost everything is out of place: "The underside of the skull has moved up, the nasal has moved back and squashed the parietals. . . ." Gary said. "It's like going back to square one for fig-

uring out the basic parts." He picked up the *Inia* specimen—the skull of an animal who had lived in the Río Apure in Venezuela in 1968—and though agnostic, invoked the deity. "God," he said softly. "Look at this weird animal. . . ." The top of the skull is crunched together like a knob, and the jaw is fused, forming a Y, with pointed, single-cusped teeth along the stem and complex crunching teeth arrayed in two rows along each branch.

The story that shaped these skulls lay embedded for millennia in the shallow sediments of extinct rivers and inland seas in Pakistan and Egypt. In 1978 in Pakistan, University of Michigan paleontologist Philip Gingerich found, in sediments of a vanished river that had once emptied into a shallow sea, the 50-million-year-old skull of the oldest known whale—*Pakicetus*. The skull had the long jaws of later whales, but its teeth were distinctive: its massive, triangular, multicusped molars were like those of a group of now-extinct mammals called meso-nychids.

The mesonychids, as Gary explained, were what today is considered an oxymoron. They were flesh-eating ungulates, carnivores with hooves. They arose more than 60 million years ago, just after the dinosaurs disappeared, branching off from the animals who would later become modern cattle, pigs, and deer. About 35 million years ago, the mesonychids vanished from land. But Gingerich believes that their living descendants swim in the world's seas and rivers, as whales and dolphins.

Fifty million years ago, *Pakicetus* was still a changeling—a creature not fully terrestrial, nor fully aquatic, either. It had to keep its head above the water to hear well. Directional hearing in water requires that the left and right bones known as tympanic bullae be isolated in separate, foam-filled sinuses. These structures are found in the skulls of modern whales and dolphins but were absent in *Pakicetus*. And the structure of the bones of the ear suggests the animal could not dive to great depths—the eardrums would have popped. *Pakicetus* may have still spent much time on land, like otters or seals; like them, it may well have mated and given birth on land.

From the size of the skull, Gingerich estimates that *Pakicetus*'s body may have stretched fifty feet long, but that is just a guess. None of the rest of the skeleton was ever found.

But in December of 1989, Gingerich and his team made the discovery that paleontologists had awaited for a century. In the trackless desert of Egypt's Zeuglodon Valley, his team found a whale with legs.

The creature, *Basilosaurus isis,* had been known since 1835 (when it was taxonomically misclassed and named "King of Reptiles"). Dozens of incomplete skeletons of this fifty-foot, eel-shaped creature had been unearthed from this valley, which 40 million years earlier had been a shallow bay. Gingerich hadn't been looking for legs, but here they were: not only the thighbone, or femur, but also the knee joint, and finally, the ankles, feet, and four toes. The shape of the thighbone showed flat areas, where large muscles attached at both ends. Here was proof: whales had once walked.

Then in 1993 came another astonishing discovery. From riverbed sediments near where *Pakicetus* was found, but perhaps a million years younger, a team led by Hans Thewissen of Northeastern Ohio Universities unearthed the impossible: a whale with hooves. The finding was so astonishing that it prompted Annalisa Berta, a San Diego biologist, to ask, in an article in the journal *Science,* "What Is a Whale?" Gingerich and Thewissen's discoveries, she argued, challenged the very definition of the mammalian order Cetacea, to which the whales and dolphins belong.

Thewissen named his fossil *Ambulocetus natans*—the Walking Whale That Swims. Its bones were discovered among beds of fossil mollusks, showing it had lived in a shallow sea. The size of the vertebrae, ribs, and limbs shows it was probably as big as a male sea lion, weighing perhaps 660 pounds. Its front legs had become flippers. But the four-toed back feet were definitely hoofed, almost like a pig's.

What had it looked like? How had it moved? What did it eat? "*Ambulocetus* looks like a mammal version of a crocodile," Gary said, flying instantly back to the early Eocene. "It would lie in wait in the shallow water for something to eat. It would have lurked in the mouths of rivers. In addition to fish, it may have eaten other mammals when they would come to the riverbank to drink."

What would the Walking Whale have seen as it looked up out of the water? What victims might it have grabbed? "Indo-Pakistan was a separate continent at the time, not yet having crunched into the underside of Asia to throw up the Himalayas and the Tibetan plateau. There were various hoofed mammals, scarcely larger than a terrier," Gary answered. "And there were doglike flesh-eaters, and the meat-eating ungulates called mesonychids, and early rodents and primates. . . ." There were fantastic creatures, whose names sound like the poetry of another planet: hyaenodontids, anthracobunids, tillodonts. The plant-eating tillodonts looked like large rodents, and as for the anthracobunids—"we don't know what the damn things looked like," said Gary, "almost all we know are teeth and little pieces of jaws." Fossils of short-legged, hoofed animals called anthracotheres, with low-crowned chewing teeth, appear in Asian sediments shortly thereafter; their living descendants are the hippos, explained Gary, and molecular evidence now suggests that whales may be more closely related to hippos than to the extinct meat-eaters with hooves. "So perhaps," he suggested, "early whales were munching their own more terrestrial relatives!"

But soon the picture fades. Whales evolved in the Eocene, but they looked nothing like the whales and dolphins of today. What came next? No one knows. For there is a gap in the fossil record of good whale specimens for more than 10 million years. That gap lasted for most of the Oligocene epoch. Only at the very end of the epoch do whales again surface from the sediments of ancient seabeds. And when they do, they are again transformed. These early whales looked much like the primitive *Inia* of the Amazon today: creatures completely lacking hind limbs and with broad, flippered hands, adapted perfectly to life in shallow waters.

Long after we had returned home, I asked Gary if he saw any similarity between his stories of the bufeo's ancient ancestors and Don Jorge's stories of how bufeos behave today. He laughed. "Scientists like to believe they live a wholly different world than storytelling," he said. "When we tell a story we like to call it hypothesis formation." Unlike the storyteller, the scientist is ethically bound to try to disprove his hypothesis—an important distinction, Gary stressed. But yes, he conceded; there were similarities, too.

"As a child, I used to think of the curious juxtaposition of dragons versus dinosaurs," Gary told me. "Neither is part of modern existence. Both are made of fragments. We are both filling in. In science, we have the fossils we are fleshing out; and the people are still fleshing out their stories, too."

In science, as in mythology, humans seek the connections in the world—the links between reptiles and birds, between land-dwelling mammals and water-dwelling dolphins—and, for that matter, between a prayerful planting and a plentiful harvest, between acts of good and evil and rewards and punishments. *Archaeopteryx* supplies a link around which paleontologists have built a story; *Basilosaurus* supplies another.

But the Cobra Grande, the Mapinguary, the water jaguar, are also links. Whether or not these creatures now dwell among us in flesh and blood, or whether they dwell in our memories or imaginations, they provide connections, too: our behavior has consequences. To risk enraging these ancient monsters invites grave danger, so people avoid areas where they have reportedly been seen. The stories, in effect, create moving "reserves" where people do not hunt or fish. Such fears, in fact, protect the bufeo, which is not hunted. I remembered what Don Jorge had told us before Gary and I set out to look for bufeos: *"It is good to fear it. It is always good to fear it."*

"Much of the lore surrounding Amazonian waters serves to conserve resources," Smith points out in *The Enchanted Amazon Rain Forest.* "I do not wish to imply that indigenous peoples are always in harmony with nature"—archeological records show that certain areas of the American tropics suffered severe soil erosion many hundreds of years before Europeans arrived, because of indigenous peoples' unwise farming practices. At one time, the Amazon floodplain was densely settled: some chiefdoms extended more than sixty miles inland. Estimates of human populations in Amazonia around 1500 range from 1 million to 6 million people—the latter figure approaching the density of today. "But the fact remains," Smith points out, "although pre-contact aboriginal populations were dense, they did not trigger massive destruction of natural resources. Cultural checks were evidently in play to prevent abuse of fish, game and other natural resources."

But are the stories the people tell today of these fantastic creatures really true?

As Gary would argue, the stories do not conform to scientific facts the way that fossils do. The truth of scientific stories can be tested, he points out. But so can the people's stories about the Cobra Grande, about Curupira, about dancing dolphins. "Veracity," said the British biologist Thomas Huxley, who bravely championed Darwin's theory of evolution, "is the heart of morality." The word "truth" comes from the Old English *treowth,* which means "loyalty." Perhaps these stories persist because they honor a loyalty, a morality, that dictates how we should behave toward other creatures. Curupira yokes us to this truth: killing too many peccaries is a sin against the human community, which depends on a regular supply of peccaries for food. But Curupira's story also reminds us of a deeper truth: greed, he tells us, can suck the very humanity from our souls.

The river people's stories tell us a truth as real as buried fossils. Don Jorge's story and Gary's were, in fact, mirror images: The shape-shifting dolphins come to us from different ends of the spectrum of time. One story tells the whales lived on land, transformed, and slipped into the water; in the other, the whales arise from the water, transform, and walk on the land. The stories are equally fantastic. And both story-tellers assure you: This isn't just a story. It's really true.

There was a time when whales walked. There was a time when they may have made love on the land. And even today, when the Amazon's waters drain low, sometimes the bufeos, the most primitive of the living whales, are left in the shallows at the water's edge. At times, on their great winglike flippers, each with five fingers inside like ours, they reenact the turning point of their evolution, perhaps 50 million years ago: still, they can crawl upon the land, sometimes for many dozens of yards, to find their way back to the water once again.

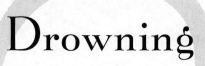

Drowning

"THE DOLPHINS DO MANY THINGS IN THE JUNGLE," MOISES began one afternoon. He told us the story of the Vasquez family.

They had a beautiful daughter, Doselina. One day fourteen years ago she was sitting on the porch of the house when a man greeted her—a man she didn't know. He said he was from a neighboring village. "Where's your canoe?" she asked. He said he had walked to the house.

"What did he look like?" I asked.

"His face was light, pink like American people," Moises said, "and he had black shoes, a beautiful watch, and a very good flashlight."

Naturally, the girl fell in love with him. The couple planned to marry in three months. But she was afraid to tell her father, who was very protective of his girl. The couple met only at night, when her lover would come to the house and slip into her bed. But early one morning, as the young man was sneaking away, the father woke up and shot him.

No one ever found a body. But the next day the father found a pink dolphin dead on the beach. The girl never saw her suitor again.

A year later, though, she was out on the river when a strange wind came out of nowhere and capsized her canoe. Her body, too, was lost. But the father had a dream that night: "I wanted to marry your daughter," a pale man told him in the dream, "but you killed me. And now your daughter and I are finally together."

MAMIRAUÁ: CALF OF THE MANATEE

I woke Dianne at 11 P.M. with a line from a nursery rhyme: "Hickory, dickory dock," I recited into the darkness.

"A mouse ran up the clock," Dianne answered sleepily.

"A rat, actually," I said. "And no clock—my legs."

For the third time in five minutes, as I lay in the top bunk bed at the Projecto Mamirauá office, the animal had scampered over the sheets covering my feet, ankles, and shins, and would have continued north had I not knocked it off the bed. But despite my efforts, it seemed determined to return.

I'd noticed the rat in the room when I'd piled our gear in there earlier, but had decided not to mention it to Dianne. The facility in which we found ourselves sleeping that night also hosted three manatees, a giant Amazon river otter, and a white uakari monkey—all orphans rescued by Project staff. Where there is animal feed, I knew, there are usually rats. But I like rats. I had assumed it wouldn't bother us. I certainly did not expect to find it in my bed.

Minutes before, I'd lain thinking that this day, nearly over, at least could not get worse. Wrong again.

Dianne and I had flown to Manaus several days earlier. Now it was the end of August, and I was still obsessed with the idea of following the dolphins. It seemed there was only one way to do this: to follow Vera's radio-tagged animals at Mamirauá, four hundred miles to the west. Mamirauá is the largest flooded forest reserve in the world, 2.7 million acres, at the junction of the black-water Japurá River and the white-water section of the upper Amazon, the Solimões.

We'd been unable to forewarn Vera of our coming. For weeks, we couldn't raise her by fax, phone, or e-mail. It was the beginning of the dry season—a good time for tracking dolphins because they are not so spread out. But this was also an El Niño year—a periodic warming of the ocean surface off the western coast of South America, with widespread weather effects—and now it was so dry that hydropower was limited and power outages common. No wonder we couldn't reach Vera. We'd decided to go to Brazil anyway. "At least we have a better chance of fol-lowing dolphins in Brazil than in the States," I'd said to Dianne.

We arrived at INPA to find Vera beset by a conference. Not only couldn't she accompany us to Mamirauá; she didn't even know for sure where her telemetry receiver was, she had packed in such a hurry. She'd been in Mamirauá just the week before, unsuccessfully hunting for her radio-tagged dolphins. Only three of the nine transmitters she had originally affixed to dolphins, attaching them by nylon pins through the dorsal ridge, seemed to be working. The others' batteries had died or the transmitters had fallen off. But in eight days of search-ing, Vera hadn't been able to pick up a signal. She wasn't even sure these three were still working. The seven months they had so far func-tioned was already a record for a transmitter attached to a dolphin.

In addition to the data she takes from her boat, three ninety-six-foot telemetry towers, strategically placed at the entrances to big lakes, au-tomatically record any "hits" from the radio-tagged dolphins who pass by them. Though she had not analyzed all the data from her seven months of tracking, so far it strongly suggested that although the dol-

Mamirauá Reserve

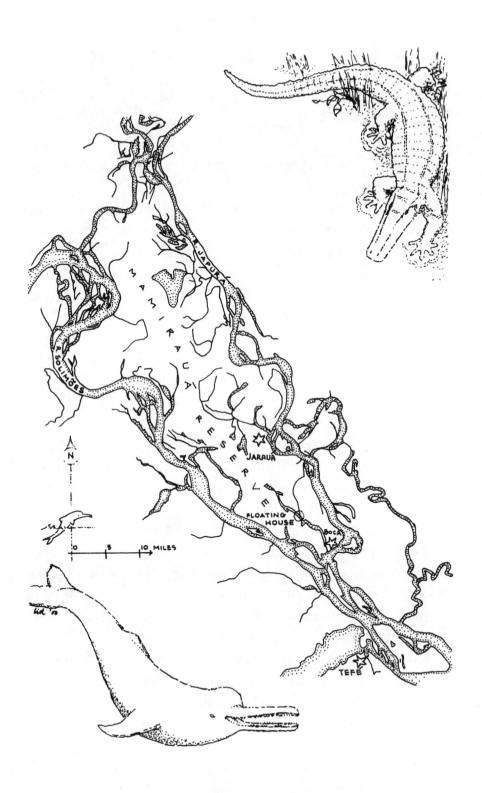

phins can swim nearly nine miles in an hour, they usually travel only about six miles a day, and they remain all year in the Mamirauá system—except for one case, which I wasn't sure Vera believed. At the conference we had attended in Florida, a German researcher, Thomas Henningsen, had reported seeing a dolphin with a radio transmitter on its dorsal ridge a thousand miles away from Mamirauá at Pacaya-Samiria reserve in Peru.

The tower-tracking made Vera's job easier, but not effortless. "You will see the towers I have to climb," she said, laughing. "To be a biologist, you do crazy things. You have to fix engines, you have to climb thirty meters, only to study the animals." One of the towers, Vera said, had been colonized by Africanized bees. Nonetheless, dressed in a beekeeper's outfit (an expense she knew would raise eyebrows at INPA—a beekeeper's suit for a *dolphin* project?) Vera had climbed up the tower in June to download the data from the receiver—which, she found, was full of honey. Another of the towers was actually a platform atop a tree inhabited by hundreds of large, biting ants. "The ants, when you go up, are not so angry," she said. "But when you come down, they try to bite. I just push them off. I do not kill." Vera mentioned the tree offered a good vantage point from which to observe dolphins who passed by. "Maybe you should go up there," she suggested. Dianne and I looked at each other and thought, Maybe not.

Vera gave us a phone number for Márcio Ayres, the Brazilian scientist whose studies of the rare white uakari, a white-coated version of the red uakaris we had met at Roxanne's camp, had convinced Brazil to establish the reserve, since the white uakari lives nowhere else in the world. Back at our room at the Hotel Monaco, I tried to phone him. Seven times I tried the number with the same result: I would dial 9 for reception, ask for the operator (who always answered with great surprise), and give her the phone number in Portuguese and hang up. She would ring our room, I would pick up the telephone—and hear no sound whatsoever.

I walked down to the desk to report the problem to the English-speaking receptionist.

"Why no sound?" I asked.

"There is a problem," he answered.

"What is the problem?" I asked.

"I think," he replied, "that it is the telephone."

On the eighth try, we reached a voice: *"Oi,"* I said, the Brazilian greeting, and was cut off. On the ninth attempt, I tried English: "Hello." Cut off. Tenth try: I reached Dr. Ayres, introduced myself and Dianne, and told him what we were doing. "Is it possible," I began to ask—and we were cut off. On the eleventh try, the biologist answered without wasting time on a greeting: "Yes, yes, you can come." We booked our flight to Tefé, a river town of 35,000 people, twenty-five miles from the reserve.

Our luck seemed to change when we arrived in Tefé. At the Projecto Mamirauá office, two friendly, dark-haired women in their early thirties, Miriam Marmontel and Andrea Piris, a manatee researcher and forester respectively, welcomed us in perfect English and generously made all the arrangements: a Project speedboat would take us to the largest of the six floating houses in the reserve. We would have two boatman-guides to help us—both strong, knowledgeable local men, and both named Antonio. In a second speedboat, Miriam would accompany us to the reserve. And, to my astonished delight, she agreed to loan us the telemetry she uses for her manatees for our first three days to try to track the dolphins.

Everyone at the Project office seemed to know about our expedition. Right before we left, the office manager, a handsome Brazilian named Luciano, winked at us and said with a sly smile, *"Cuidado com o boto."* ("Be careful with those botos.") "He's warning you not to let them take you away," Miriam said.

Andrea accompanied us to the *supermercado* to buy provisions for ourselves and for the Antonios. What would the men like to eat? I asked.

"First of all, you must get four frozen chickens," Andrea announced. Of course.

We also bought huge sacks of farina, rice, and dried beans, fresh potatoes, tomatoes, onions, and eggs, tins of beef and sardines and tuna, cooking oil, salt, coffee, cheese, bread, crackers, powdered milk, spaghetti, tomato sauce, oranges, limes, apples, and a powdered-drink mix (which proved loathsome) to supplement the supplies of powdered Gatorade and freeze-dried food we had brought from the States.

We loaded our gear into the two 24-horsepower speedboats at Tefé Lake that afternoon and set off for the two-story floating house that would be our base.

Within minutes, we saw our first boto: a huge pink fin rose and fell, rose and fell, before us just as we were turning left from Tefé Lake into the channel that leads to the Solimões. Here, as at the Meeting of the Waters, white and black waters meet, and the dolphins, Miriam told us, love to hunt here.

And then, another fin—a dark one—and another, a gray. Our hearts leapt. At least three, perhaps five, dolphins surfaced around us, blowing. Miriam said she thought some of them had respiratory problems, they breathed so loud.

Just after these sightings, however, one of the boats—mine— abruptly stopped. Antonio tried to fix it, but soon announced it was *"avariou-se"*—broken-down. We would have to be towed back to the Mamirauá Project boathouse. The Project, explained Miriam, was impressively equipped with five houseboats, for remote and overnight trips, and twelve speedboats. "But of course," she said, "they don't all work."

We would have to get a new boat tomorrow, she said. Dianne and I

A rat slain by Dianne.

could sleep at the Project office. We stowed our groceries, in flimsy plastic bags that were already splitting, at the boathouse, where I was sure they would be eaten by rats. We carried the four precious frozen chickens back with us to the Project office, which had a freezer, and got ready for bed.

Dianne took a sleeping pill. But before falling asleep, she had, as usual, spread much of the voluminous contents of her bags over our small room for a final inspection. This is why my 11 P.M. rat report so roused her: the rat might run over her *clothes*.

She launched out of bed to help herd the animal into the bathroom.

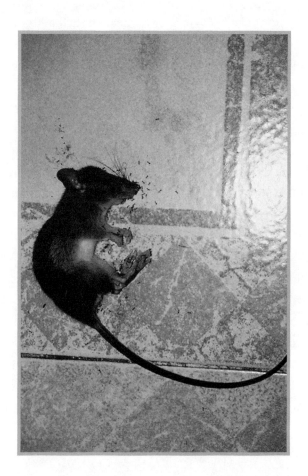

Unfortunately, the door wouldn't close, a fact we had earlier noticed as clouds of mosquitoes rose from the shower drain and poured into our sleeping quarters. Maybe the rat would vanish down the drain. We waited a minute, and then Dianne allowed the door to spring open a crack. The rat saw its chance to rush out. With admirable speed and force, Dianne pulled the door shut. When the door sprang open again, the rat lay motionless on its side. Blood oozed from its snout.

"You killed it!" I shrieked, horrified.

"I killed the bastard!" she exulted.

"You fucking *killed* it!" I cried ungratefully.

"I killed it! I killed it!" she shouted in glee, and then loosed a laugh like a pirate.

We went back to bed, leaving the rat dead in the bathroom.

Five minutes later, the scuttling began.

"He's not dead," came Dianne's voice, disappointed, in the dark.

But when I turned on my flashlight, I saw this was a different rat: a much bigger rat, with prominent testicles. It was peering at me from the foot of my bunk.

We herded this one, too, into the bathroom. This time we tied the door shut with a bandanna stretching to the knob of a nearby dresser. I took one of Dianne's sleeping pills. If more rats came into my bed that night, at least they didn't wake me.

The next day, equipped with a new speedboat, we arrived within the hour at a little village, Vila Alencar, to pick up the second Antonio at his wooden, tin-roofed house there. Next we stopped at the floating house at the intersection of two rivers, a place called Boca, which means "mouth," to say hello to the gray-haired watchman, Joaquim. Miriam and the Antonios chatted with him perhaps forty-five minutes as the skies grew darker and darker.

"*Chovendo?*" I asked. ("Rain?")

"*Não, não,*" chorused the Antonios.

"Should we leave early and avoid the rain?" I asked Miriam, unconvinced.

"No—they say it's not going to rain."

"My poncho is in your boat somewhere—should I dig it out?" Dianne asked me.

"No," I assured her. "They say it's not going to rain."

We got in our boats and pulled away from Boca. Almost immediately, the downpour began.

Rain pounded all around us—big opalescent drops, heavy as stones. Thunder and lightning exploded everywhere, a battle of water and fire. I remembered something Vera had said the night we'd dined at the open-air restaurant, as lightning crashed all around: "In my whole life before coming to the Amazon, no one died of lightning. But here it is not rare. I know of three at INPA who have died."

The rain stung my skin like buckshot. I averted my face and stared at the floor of the boat. Helplessly, I watched it fill with a shallow lake of rainwater—disintegrating the flimsy plastic grocery bags, reducing our bread to sodden Kleenex, turning our crackers to paste, thawing the frozen chickens, and drowning our luggage—luggage in which Dianne's poncho was irretrievably embedded.

Twenty minutes later, we staggered soddenly into our floating house. The chickens, completely thawed, dripped blood over everything. Dianne's lips were blue. Miriam looked miserable. The Antonios seemed frozen. I had one thought: get hot drinks into them, *now*. I raced into the little kitchen to light the gas stove. My hair, hands, and clothing dripped water onto the matches, extinguishing one after another. Finally, one stayed lit long enough for me to turn on the burner. The gas blew the match out.

"You go through a box of matches every time you light it," said a voice with a British accent behind me. I turned around and noticed for the first time, and to my horror, that our floating house was already inhabited.

Paul Sterry, forty-three, a renowned nature photographer, gallantly lit the stove for me. Young Lee Morgan was a student at Royal Holloway College, University of London, studying the behavior of ringed kingfishers. Andrew Cleve, elegant and graying, was the warden of the Bramley Frith Study Centre near Basingstoke in the U.K. and the author of more than twenty-five books on natural history and biology; for his service to the environment, he had been given a medal by the queen. Peter Henderson, forty-four, a professor of evolutionary biology at the Animal Behavior Research Group of University of Oxford, had been studying the fishes here since the reserve's inception seven years ago; he probably knew more about the dolphins' underwater environment than anyone alive.

As we sipped hot coffee around the big dining table of the spacious central room, Peter told us about his early days here. "When I first saw this place, I thought, How on earth can I possibly work here?" Though blond and blue-eyed, he reminded me of Gary, with his encyclopedic knowledge and gift for storytelling. "It seemed like hell at the time."

He had lived with Márcio Ayres, the uakari researcher. They had felt marooned, living on houseboats and an abandoned floating house, spending weeks without setting foot on land. Márcio could hardly find his uakaris; Peter had no idea how to sample the fishes. The place swarmed with mosquitoes and biting horseflies called mutucas. The only diversion was to visit Tefé, back then (before the government stationed 5,000 soldiers there, to guard against the dreaded "internationalization of the Amazon") a fishing town with only four cars and no restaurants, where dogs slept in the streets and people slept on their porches. "We wondered," Peter said, "whether we'd made the biggest mistakes of our lives."

Twice, Peter was nearly killed while working here—both times by creatures he never thought to fear. One day a sloth fell out of a tree and nearly hit him. It looked him in the eye, rolled over, and died. Another time, a plant or insect poisoned him. To this day, he doesn't know what touched or stung him—only that his eyes swelled shut, his hands were paralyzed, his scalp went stiff, and finally, while vomiting, he fainted. When he regained consciousness, he found himself in a clinic in Tefé, breathing oxygen from a tank marked "for industrial use only."

Yet he comes back every year, often with eager colleagues. For him, the place exerts the primal pull of a sea. In fact, many of the creatures here would seem more at home in an ocean than a lake: There are spotted freshwater stingrays with poisonous tails, oceanic fishes like sole, goby, and herring, four species of crab, and of course, dolphins. "It's as if the scale of the freshwater is so great it almost becomes like an ocean," Peter said, his blue eyes shining.

Mamirauá is a place of oceanic scale. It is half the size of the country of Belize, and possibly the world's richest aquatic system, with 499 lakes in the central study area alone. Besides the largest protected flooded forest on earth, Mamirauá is also Brazil's largest conservation experiment: in 1996, six years after the state governor created the Mamirauá Ecological Station at Márcio Ayres's urging, the area was proclaimed the nation's first sustainable development reserve. Instead of being evicted, the 2,500 people who live in and around the reserve were not only permitted to stay and use its resources, but to serve as its guardians—a bold attempt at protecting wild land and helping people at the same time. It was an experiment so novel that the creation of the reserve forced Brazil to alter national conservation legislation: never before were local residents allowed to fish and hunt in areas declared reserves.

Mamirauá is a huge natural experiment as well. In the sealike lakes and along the snaking rivers, forest and water merge, giving birth to hybrid creatures that seem to defy possibility. There are fish that nest in trees, fish that hunt in air, and fish that incubate as eggs inside the father's mouth. Sloths swim like athletes, their long, claw-tipped arms crawling through the rivers. Whole meadows of grasses flower, floating, atop the water; and 150 shape-shifting pink dolphins cast nets of sonar as they hunt in submerged treetops.

Peter has found at Mamirauá an unrivaled laboratory of evolution.

In a place like this, where the world as we know it was born, the world still re-creates itself anew. "These flooded forests are the habitat from which terrestrial groups evolved," he explained, "yet, though this habitat has been around for millions of years, nothing here is very permanent." The water rises and falls, carves new channels, builds new islands. Half the floodplain here has been entirely resculpted in the past millennium. The lakes are actually only a few hundred years old. Dredges have unearthed shards of Indian pottery from the lake bottoms, remnants of a thriving Omagua civilization whose language, now gone, gave this place its name: "Mamiraua" means "Calf of the Manatee." Everything here is in transition, in the process of becoming something else. Trees evolve into weeds, like the cecropias, and weeds into trees: the ephemeral herb along the banks, *Acoitis aequatorialis,* belongs to the plant family Melastomaceae, almost all members of which are trees. Creatures assume an astonishing spectrum of forms to exploit new niches: Lily pads grow into three-foot-round giants as well as miniatures just .04 inch long, only root and leaf. The fishes have evolved in staggering array. Peter and his colleagues have catalogued 312 species here, 80 species of which they have caught from the front porch of this house. By comparison, all the species of fish in Peter's native England number only 30.

Even now, as we watched the rain clear as we sat around the dining table of the floating house's spacious central room, the land was changing almost before our eyes. No longer was this the luxuriant, brimming wet season, as we had witnessed at the Meeting of the Waters and along the Tahuayo. The water here had dropped twelve feet in the three months since May, Peter told us. Along the riverbank across from our front door, we could see where the ebbing river revealed holes the size of croquet balls—dens excavated by catfish in the wet season. Several species of fish here dig dens: the giant pirarucu, ancient fish whose bony tongues are covered with teeth, use their chins, fins, and mouths to gouge nest holes in the mud of shallow waters, and male and female together guard the hole, like birds. Another fish, the goby, nests in sponges that grow in the curved leaves of submerged trees.

But the season of nesting was long over. The dry season is the killing season. "Everything that's in the water here very quickly goes into a body," Peter said. "Here is a sort of killing zone, where the feeding is going on intensively." Piranhas evolved in the Amazon's flooded forests, Peter told us. "They are a great innovation," he said. They are mainly scavengers, like hyenas, he told us, and generally don't attack healthy animals. Like hyenas, piranhas are social animals, traveling together in groups of individuals who are probably related, and vocalizing all the time to stay in contact. "Piranhas have a stinking reputation in Brazil," he said. "Prostitutes are called piranhas of the land. But I think we should make the piranha the symbol of the Amazon, for they are one of the fish that truly molds the ecosystem. As an animal, I find them very elegant and very impressive indeed," he said—although, he confessed, "I don't like handling them."

At the moment, the piranhas were having a field day. At high water, they have to hunt mainly at river edges, where fish may have become stranded, while in low water, the fish are so concentrated they can attack anything that moves. For the dolphins, too, this was a time of ease and plenty. At the edges of open water like that in front of our house, fish were as densely packed as five per cubic meter, as measured by Peter's sonar. "Every cast of the line, you bring up something," Peter told us—although often it was only a head, the rest of the body devoured by piranhas. Later, when we would throw our dinner scraps overboard, the waters would boil with piranhas.

Every five or six seconds, we would hear a splash. Often it was a kingfisher plunging, spear-billed, into the water. No fewer than nineteen ringed kingfishers have established territories along the 1,600 feet of bank directly across from our front door, Lee had found. Fish are so concentrated that the birds were probably not defending the fishing rights to their territories, but instead, he suspected, defending the rarer, shady perches on which they remove the spines and scales of their prey.

The splashes signaled not only birds plunging into the water; as we watched, we realized fish were also leaping *out* of it. Sometimes they leap to escape underwater predators, Peter explained. Others skim the surface like stones. Many other fish come to the surface to breathe. Some can breathe no other way. The giant electric eel, the pirambóia, or lungfish, and the giant pirarucu must surface every ten minutes to gulp air, for long ago in their evolution their gills ceased to function as organs of inspiration. Still other fish leap into the air to hunt. The lithe, smooth aruana swims near the surface with chin whiskers forward, the tops of its eyes projecting out of the water or just near the surface; the eye is divided horizontally, so it can see above and below the water at once. When it spots its prey, it erupts from the water, sometimes launching more than six feet into the air to seize beetles, spiders, and other creatures from branches, vines, and trunks. Occasionally, an aruana will take a small bird or bat perched on a limb above the water. Michael Goulding reported a case in which a large aruana, over three feet long, took two newborn sloths from their mother's arms.

Later, on the water, we would see fish surface and leap into the air beside our boat. Their scaly, prehistoric faces leered up at us, unblinking, like images from the subconscious surface in dreams. At the water's edge, we would see aruana hunting. The fish seemed to hang in the air, a long silver ribbon, before falling back into the water. And almost daily, to our amazement, dolphins would surround our boat. It seemed, at last, that the waters would finally open for us. And they did. But, as we were to find, it was not in the way we expected.

By midmorning, the skies had completely cleared, but one of our boat motors had died. Miriam needed one boat to return to Tefé in the morning, and Dianne and I needed the other. What to do? Miriam suggested we try to borrow a boat from her colleague Ronis, the way one might ask to borrow a cup of sugar from a neighbor. Ronis da Silveira, the caiman researcher, lived with his wife, Barbara, at the four-room floating house nearest to ours, ten minutes away.

Ronis had married his slender, smooth-browed sweetheart, an agronomy student, just one year ago. They were wed in a ceremony at the Manaus opera house. Then they had come here, to drink rainwater collected in a barrel, shower in water pumped from the river, and fall asleep to the weird, wet calls of birds and frogs that Barbara had never before heard. It was her first time in the Amazon jungle.

A recent issue of the international Crocodile Specialist Group newsletter reported on the couple's honeymoon: Ronis, his new bride, and his assistant Edejalma had tackled a fifteen-foot black caiman. After wrestling with them for forty minutes, the giant reptile had tired sufficiently to allow the trio to drag it beside the aluminum canoe back to the floating house at Boca. While they were affixing the transmitter to its scutes, the caiman saw an opportunity to escape. It hoisted its tail six feet out of the water, grazing Barbara's face with its horny scales. The bride's tender face, however, deflected the tail sufficiently to cause it to miss the boat when it came down. Were it not for Barbara's face, the boat almost certainly would have overturned.

A giant black caiman skull, its jaws propped open by a blue and red basketball, greeted us at their door. It had belonged to an animal thirteen feet long. "Killed by jaguar," Ronis told us; it had been one of his radio-tagged study animals. A jaguar, I thought, was probably the only other creature besides Ronis who could have subdued a caiman that size. Muscled and macho, with dark Latin eyes, the torso of a bodybuilder, thick black hair tied back in a ponytail, and a heavy beard evident even when he shaves, Ronis reminded me of the comic-book hero Conan the Barbarian. (Lee had said he looked exactly like the star of a British strip called *The Slayer.*) In fact, Ronis does read Conan the Barbarian comic books, and jokes that he does so to psyche himself up to go catch black caiman—the largest predator in the Amazon, which grows to sixteen feet, and is more numerous in Mamirauá than anywhere else in the Amazon. But in reading the comics, Ronis also pursues a more scholarly goal. Once, at a time when a visit from important American researchers was pending, Peter found Ronis poring over his comic books with great concentration. Ronis looked up from the pages at Peter. "English," Ronis said earnestly. "Must learn!"

"That's the kind of guy Ronis is," Peter had told us earlier; "he looks macho, but he's a true gentleman."

And this he proved to be: Not only did he loan us a boat, but he also promised, to our delight, to take us out one night with him looking for caimans.

But our days belonged to the dolphins, and we wanted to make the

most of the telemetry equipment. We realized, to our horror, that the equipment alone wasn't enough. We did not know the dolphins' radio frequencies. Each animal has its own; to search for any tagged individual, you essentially "dial up" your dolphin by plugging its three-digit number into a receiver box, attached with a cable to an antenna. You hold the antenna vertically, moving it in an arc, to scan for that particular signal. A kissing sound on the earphones indicates the dolphin is within range, and the sound grows louder the closer you approach.

Happily, Miriam told us, Vera's dolphin frequencies were recorded here at Mamirauá. They were marked, she said, on the receiver box at the top of the telemetry platform at the entrance to Lake Mamirauá—ninety feet up the apuí tree with the giant, biting ants.

The apuí was a stout, welcoming creature, with wide-open arms like an apple tree. Red and green painted boards nailed to its trunk provided sturdy hand- and footholds. As we climbed, following Miriam, we watched carefully for the biting ants. We saw them almost immediately: inch-long black and gray insects with prominent mandibles. "Ants!" we chorused. There were seven to ten of them every square foot or so. They did not march in columns, along a scent trail, but patrolled as if at random, which made them more difficult to avoid. If we crushed one, it would release a chemical siren summoning the rest of its clan, who would rush in bravely to attack us. This fact of ant behavior, and not the basic reverence for life, we now realized, was the reason that Vera made a point of never killing them.

Carefully, we climbed, each step choreographed not to anger the tiny beings of which we were, I realized, rather absurdly afraid. ("But they are not poisonous!" Moises would have reminded us.) Actually, I was less worried about being bitten than about my reaction to them if I were; if I let go my handholds, it would be a long, swift trip down. Of

Searching for dolphins with radiotelemetry.

this, Dianne was acutely aware. Her other fear, besides spiders, turns out to be heights. Neither of us had suspected that our dolphin expedition would feature so many fist-sized spiders and vertical ascents. "I know, I know," I said to her below me, "this wasn't in the brochure."

"We're almost there," Miriam said encouragingly. She had climbed this tower many times, to download data from her manatees' telemetry. On rare occasions, she has stood on the platform and watched manatees pass into the lake, their blunt, cloud-shaped forms floating up through the dim waters to nibble water hyacinth. "There's something about their gentleness," Miriam had said to us when we'd asked her what drew her to these unlikely creatures, "being so big and so gentle." In English, these huge, placid grazers are often called sea cows; in Portuguese, they are similarly perceived, and called *peixe-boi,* which means "fish-bull." The word is pronounced "peshy-boy," a trusting, childlike sound we loved. Certainly, it had fit the manatees we had met back at the Project office. We had watched a staffer feeding one of them, a two-year-old orphan named Boinha, a process both man and manatee obviously enjoyed. Wearing a red cap, the staffer lay on his belly on a board over the tank house, holding the outsized baby bottle for the outsized, six-foot baby. Idiosyncratically, she had nursed upside down, her square, gray tongue pressed against the roof of her mouth over the nipple. She had let him cradle her stubbly snout in his hand. As she sucked, her slitlike nostrils opened and closed, and she had shut her small eyes in total trust, as if in ecstasy.

That trust was heartbreaking. All three species of the world's manatees have been ruthlessly hunted for meat, oil, and pelts. As recently as 1950, over 38,000 Amazonian manatees, the smallest of the world's three species and the only completely vegetarian aquatic mammal in the Amazon, were hunted commercially in the state of Amazonas alone. And in Mamirauá, despite federal protection outlawing their slaughter since 1967, the killing of manatees is still sanctioned here in accordance with the reserve's unusual management plan.

We were glad the manatees had Miriam in their corner—and ours. If

we had to be climbing a tree full of biting ants, we were glad to be following in the footsteps of this strong, beautiful, competent woman.
Miriam had the muscled physique of an athlete, which she maintained
with a regular training program at the local gym. With beautiful light
brown eyes and glossy black hair, and a ring or two or even three on
every finger—mostly in the shape of dolphins (few rings were available
in the shape of manatees, she explained)—she was the sort of woman
men literally fought over. We later heard rumors about Miriam's jealous suitors that sounded like Wild West stories, involving knives and
guns.

But Miriam had other loves. "I was always in love with the sea," she
had told us back at the floating house. Growing up in the town of Pôrto
Alegre, she had read about a profession called oceanography, and had
gone to the University of Rio Grande in southern Brazil, 250 miles from
home, to study it. Few people from Pôrto Alegre ever leave their home-
town, she said, least of all young women; but Miriam is a maverick,
like Vera.

There was no program of study for marine mammals at the time. But
Miriam heard about these creatures from Argentinian researchers visiting her school. "I just couldn't believe such things existed here," she
had told us earlier at the floating house. "Manatees! And a pink dolphin—that was just unimaginable. I knew then what I wanted to study:
aquatic mammals in the Amazon, that's got to be it."

Miriam had traveled further, earning a Ph.D. at the University of
Florida at Gainesville, studying Florida's manatees. She saw her first
whales—humpbacks—on a trip to Cape Cod. She had loved America
and Americans, and treated us as if we had been personally responsible
for all she had loved and learned in our country. Perhaps this was why
she was so patient and encouraging with us; and also, she knew what it
was like to be a woman traveling in a strange land.

"Look—we're here," she announced. But as my head cleared the top
of the platform, I froze. I felt eyes on me. I turned my head slowly and
stared into the red eyes of two hoatzins sitting on a nest perhaps ten

yards away. They erected their strange orange crests and hissed at us like lizards.

Finally, we stood atop the telemetry platform in the treetop and gazed over Lake Mamirauá. From this height, the radio antenna can pick up radio signals from three miles away, Miriam explained. Every twenty seconds, the receiver scans a different VHF frequency; if it receives three signals in a row, it stays with that transmission and records it, tracking the animal's path. A solar-battery–powered microprocessor records the strength of each radio signal (giving an idea of how far away the animal is) and its direction. The black box next to it, about the size of a tool chest, was the receiver.

Gleefully, we lifted the lid on its hinges, like pirates opening a treasure chest. Inside, we saw two sets of frequencies listed: one for Vera's dolphins, one for Miriam's manatees. Eight of the nine dolphin frequencies were clearly marked. I copied them carefully: 272, 424, 439, 575, 726, 769, 871, 891. These were the combinations that would unlock the secrets that I had sought for so long. Finally—having combed rain forests and rivers, having climbed up trees and plunged into lakes, after consulting biologists and shamans—now we would be able to accomplish what we had come for. Finally, we could follow the dolphins.

Or so I thought.

THE WATERS OPEN

We didn't have much time. We had only three days to try to locate 3 radio-tagged dolphins among a population of 150. They could be anywhere in waterways that coiled and threaded for uncounted thousands of miles through forests that stretched over 2.5 million acres.

Dianne and I set out anxiously the next morning. As I plugged in the earphones and cables and erected the antenna, an Antonio steered us toward Boca. We knew we would find dolphins there, at the intersection of two waterways; and besides, over the following two days, when we joined Andrea and Miriam again, we would be heading in the opposite direction. We wanted to cover as much ground as possible.

Almost immediately, long before we even approached Boca, our boat was surrounded. Antonio cut the motor so we could watch them. The waters here were murky, and we could not see below the surface. But this time, the pink dolphins almost seemed to be showing us their bodies: we could see the dorsals clearly, as well as the tops of their heads. And unlike at the Tahuayo and at the Meeting of the Waters, here the botos often surfaced side by side, as do tucuxis, which made them easier to count.

There were eight, Antonio confirmed. It seemed, for the first time, that each one was distinctive. One of them was very large and pink. Two were gray youngsters, and another, despite the fact that all babies are supposed to be gray, was, like one we had seen in Peru, a very bright pink. One of the adult dolphins had an L-shaped scar on the dorsal ridge—one of fourteen dolphins that Vera had come to recognize by natural marks, a male named Scar to whom she had assigned the momentous Number 1. On the paper where Vera had drawn pictures of those with natural markings, the scars and blotches had looked obvious: Number 8, Meia-lua, had a half-moon–shaped chunk missing from the dorsal; Number 3, Ruffles, was marked by a series of frilled scars on the back side of his; Number 13, Riscos, had two deep notches in the skin behind the head. But Vera had told us she had spent entire afternoons among groups of dolphins that never gave her a good look at their dorsals. We were extremely lucky. Another dolphin, we noticed, had a hole in the fin—likely where the metal pin holding a transmitter or a plastic tag had fallen out. And another, Antonio recognized as Shika: she had an X on her dorsal, the first female Vera had freeze-branded. She blew at us loudly.

They approached us closely, some within ten yards of the boat. It was as if they had rushed to greet us—and perhaps they had. They surely recognized project boats. Dogs and owls, with hearing far less exquisite than that of the boto, know the sound of individual automobile engines on suburban streets. Perhaps the dolphins also knew the boats by sight, and even recognized the people normally in these boats; perhaps they realized we were strangers. We saw several of them spy-hopping to get a better look.

Their boldness amazed us, given their experience. To freeze-brand or equip a dolphin with a transmitter, the animal must be subdued with nets, hauled from the water for perhaps twenty-five minutes, and subjected to minor surgery. No dolphin has ever died from these procedures, Vera had told us, but it is surely painful and terrifying. It is frightening and dangerous for the researchers, too. The dolphins are easier to catch in the dry season, when the water is low, but this is also when the piranhas are most concentrated. "Nobody wants to jump in that water!" Vera had told us.

Besides, the big male dolphins—the only ones whose dorsal ridges are big enough to support a transmitter—will readily bite in self-defense, and, with 130 teeth strong enough to crush an armored cat-fish, are capable of inflicting wounds as bloody as a shark's. "The males, it's amazing the power in the jaws—*plash!*—like two boards crashing together," Vera had told us in Manaus. She has learned—at the suggestion of her mother-in-law, whose husband was director of the Vancouver Zoo—to bind the jaws with nylon stockings. But even with jaws bound, botos are formidably strong. Thrashing his flexible neck, once a captured male hit Vera in the head with his beak and flung her across the boat. She felt lucky the animal had not cracked her skull. Her face was badly bruised for weeks.

After tagging and freeze-branding a dolphin, Vera waits two or three weeks without attempting to follow it. But at the end of that period, when she seeks out the marked animal, she finds it is just as eager to approach her as before—just as she has never hesitated, even after her injury, to approach them. Vera had told us that sometimes she attracts botos by circling the boat in a figure eight. "They come very close, and will follow you and play for half an hour. It's a game, a game inside the lake," she'd said.

Why would the dolphins risk playing a game with our dangerous species? They already have plenty to play with. Like the Duisburg Zoo dolphins, who grabbed scrub brushes as toys and constructed bubble rings, wild dolphins also play with toys: At least one observer has reported seeing a river dolphin tossing a river turtle in the air, like a ball. Why bother playing with people? Perhaps because, unlike a toy, or a turtle hiding in its shell, people will play *with* them, and in unexpected ways; in this game, the players interact, one species's curiosity answering the other's.

The botos surrounding our boat clearly wondered about us, as we wondered about them. Could they know the curiosity was mutual? I'd thought so with Atlantic bottlenose dolphins I'd met years earlier. On a magazine assignment, I visited the Kewalo Basin Marine Mammal Laboratory in Honolulu, where researchers were testing the dolphins' ability to manipulate objects in response to an artificial language of symbols and sounds, to test whether they can understand both vocabulary and syntax (which the research now clearly affirms). I'd unthinkingly greeted the first dolphin I met as I would have a person: I waved my right hand at the fourteen-year-old female dolphin named Akeakamai, "Lover of Wisdom" in Hawaiian. To my surprise, she had waved back at me with her mirroring flipper. I held my leg up to the aquarium glass, and she had kicked her tail fluke in response. And when I would stand tankside, she and her female companion, Phoenix, would often rest their chins on the rail, out of the water, standing erect as I was, their tails on the floor of the tank. This was, I realized, a sort of conversation. It may have been a simple one: "Here I am!" and "Here *I* am!" But the nature of their response—to mimic my movements—implied the possibility of a deeper understanding. Despite our enormous physical differences, perhaps they recognized the sameness that we shared, a cognitive kinship spanning 90 million years of divergent ancestries. Inside their flippers, dolphins and whales have five fingers, as we do; and inside their heads, they may suspect that we and they understand the world, and delight in it, in a similar way.

The botos surrounded us for forty-five minutes. We made 135 observations of them—and they, perhaps, as many of us—before we went our separate ways. We continued toward Boca, hoping to locate the radio-tagged dolphins. Already, though it was still early morning, the sun felt white-hot on our skins, the air a pounding haze. As Dianne rolled up her shorts and sleeves to enhance her California tan, I could feel my skin burning beneath my long-sleeved shirt. My ears clogged with sweat welling beneath the rubber earphones. I swung the antenna slowly, its arc my question mark, and strained to hear the answer. Still no signal.

But within ten minutes, we were surrounded by a new group: an adult mottled gray with pink, a dark gray youngster, a dark gray adult, one with a pink stripe on the back, a very large, very pink adult, and several smaller grays. *"Mais que dez!"* exclaimed Antonio. ("More than ten!") We couldn't believe it; never before had we found two large groups in such a short span of time. Seldom had published researchers, either. The literature reports that river dolphins are usually seen alone, or just one mother with her child; only one study reported seeing groups of dolphins more often than solitary individuals—researchers studying botos at the Río Apure of Venezuela, who most frequently encountered botos in aggregations of two to seven.

Within the next half hour, more dolphins came to us—there were now at least fifteen. We felt nearly certain they included members of the first group we had encountered. We had set out to follow them; and now they, instead, were following *us.*

Did they understand that we had come to see them? I wished I knew a gesture with which to greet them, as I had with Akeakamai. To wave seemed inappropriate. Instead, I knocked on the sides of the boat. I thought perhaps they'd simply rise in response: "What's that? Let's see." But to my surprise, one chose to *answer* my sound with a sound: a loud, big bubble, a spoken explosion. I knocked again, and our conversation continued. We received six loud bursts of bubbles at lengthening intervals, all within five minutes. We named these "bubble bombs"; they were quite different from the champagnelike effusion of airy pearls the dolphins had cast around our boat that moonlit night in Peru. Much later, in a clear-water tank with glass sides, we would observe North America's only living captive boto, an elderly male named Chuckles, producing similar bubble bombs at the Pittsburgh Zoo. Unlike the delicate bubble rings Wolfgang Gewalt had observed at the Duisburg Zoo, which were released from the side of the mouth, Chuckles's bubble bombs were like giant burps expelled vehemently from the center of his gaping beak. And while the Duisburg dolphins made their bubble rings as toys, Chuckles's bubble burps were clearly produced for a different purpose—to evoke a response, perhaps to instigate a conversation. Visitors always reacted. Children shrieked, laughed, and ran; adults gasped and pointed. To Chuckles, of course, the "zoo" was the visitors outside his tank, and he seemed to enjoy provoking them into doing something interesting for him to watch.

As the wild botos bombed us with bubbles, I laughed with delight; Dianne, however, was not amused. She was trying, desperately, to photograph the animals, and all she ever got was bubbles. "They are definitely playing tricks on me," she said, "and now they're laughing." Perhaps they were. Cetaceans do seem to enjoy a good joke. I remembered an account of beluga whales who lived next to the dolphin tank at the John G. Shedd Aquarium in Chicago some years ago. As part of an experiment, a Pacific white-sided dolphin there named Kri had been trained to stick her snout through a ring and hold the position until a low-frequency whistle signaled her to swim away. When Kri began breaking her routine, the researchers were mystified—until they discovered the belugas next door, excellent mimics, were precisely duplicating the tone, duping the dolphin, baffling the scientists, and foiling the experiment. "You could almost picture the belugas laughing in the next tank," their trainer had said.

After six bubble bombs, the dolphins resumed their usual game of hide-and-seek: one would pop up, blow, and we would twirl to watch it vanish; two would surface on the opposite side, we'd cry "Look!" and they would sink. It reminded me of the game I had played as a child at the swimming pool: One child, eyes shut, cries "Marco!" and the others answer "Polo!" Trying to follow the sound, Marco tries to tag his tormentors, but after crying "Polo" the others dive or swim away, just out of reach. The botos, diving and blowing, were playing Marco Polo, and we were "it."

This, of course, was not what we had come for. We were supposed to be working with the telemetry, in pursuit of a single electronic answer to a simple question: "Where are you?" "I am here." Nonetheless, I put down the telemetry and unplugged the earphones, seduced by the pleasure of their game. I lost myself in their play, and let their motion flood my senses: wet skin gliding against wet skin, the kiss of air, wind and sun on arched backs, the embrace of the cool water. Over and over, they surfaced and plunged, sliding, timeless and weightless, between water and air. No wonder botos enter human story as lovers; they glide through the elements the way lovers slide through one another's bodies, a tension of tenderness and hunger, poised on the threshold of joy. As the sunlight poured over us, heavy as honey, sweat drenched my hair, my bra, my shirt, my socks, my shoes. Sweat ran into my eyes and mouth and ears. But I never noticed until they left us, and then my mouth would water as if hungry, and I would feel tears stinging my eyes.

All morning, and later that afternoon, groups of dolphins surrounded our boat, rushing to us as if to embrace, and then swam and dove and returned again—a flirting game of hide-and-seek. Of course, we lost every time. Dianne never got a photo; I never got a "hit" on the telemetry. Yet we felt as if the waters had opened to us—and then swallowed us whole.

One of the dolphins, a dark gray one, surfaced so close to our boat I could have touched her. She opened the top of her head to us. I stared down her blowhole, an intimate, mysterious abyss of life, and inhaled her moist breath.

We were supposed to meet up with Miriam and Andrea at 11 A.M. at Boca the next day. The morning of the meeting, we searched again with the telemetry; again, though we found many dolphins, none answered my questing antenna with its electronic kiss.

Now we waited at Joaquim's floating house, but at 1 P.M. there was still no sign of Miriam and Andrea. Had we mixed up the time or place of the meeting? Had our friends met with trouble? Had they forgotten us? Of course, there was no way for us to know. We waited, afloat in the middle of nowhere, feeling isolated and cut off.

At one-forty, a speedboat arrived. It was Ronis. "You know Princess Diana?" he asked. "She die," he announced. "Have cigarette?"

Dianne gave him a Newport in exchange for this news, and, his English momentarily exhausted, he departed.

. . .

Miriam and Andrea arrived on the big houseboat, the *Uakari*, some-time after two. They had a list of errands to accomplish before we could head north to Jarauá: they had to deliver a new gas-powered refrigera-tor to Ronis and Barbara; they had to rescue a floating telemetry tower, which had run aground. On the way, Dianne and Andrea, Miriam and I sat on the bright yellow roof of the *Uakari*. Miriam and I swapped the telemetry equipment, monitoring the channels for dolphins and mana-tees.

"We used to think the pink dolphins migrated, and the manatees didn't," Miriam told us as I held the antenna aloft on a pole, to increase its range. "But the manatees are the migrants, and the travel they do turns out to be pretty impressive."

As Dianne and I scanned for dolphins, Miriam told us about her manatees. She'd begun her telemetry project in 1994, and by now had attached radio transmitters to six of them, by means of a belt around their paddle-shaped tails. She discovered that Mamirauá's manatees migrate more than sixty miles out of the reserve. They begin to leave the area as the dry season begins—they were moving now. But they don't leave en masse. Manatees are basically solitary animals, and when they migrate, they leave alone, at different times, choosing dif-ferent routes to different places. An individual may even take a differ-ent route each season; one of her study animals, a male named Zé Taboca, spent one dry season in the Solimões, and stayed another dry season at Lake Mamirauá. "It's intriguing," Miriam said. "They spend six months at the headwaters of the Solimões, and then two months in the busy part of the river, with all those fishermen and the dangerous boat traffic, and they are generally thought to be very shy." How they know when to leave, how they decide where to go, and what they do

Manatee researcher Miriam Marmontel pilots a speedboat to
the floating house. (Photograph by Sy Montgomery)

when they get there, no one yet knows. But they have their reasons, and they must be good ones. "Two weeks ago, when we were tracking them, the water was going down, and then suddenly, the water went *up*—and the manatees moved. The manatees knew before the people knew," Miriam said.

Yet people generally don't consider manatees very bright. "People say they are *cretinoso,*" Miriam told us, shaking her head, "just because they are slow."

Researchers who have examined the brains of manatees long ago decided the big, gentle animals were dumb. The brain is relatively small

for its body size, and the surface is very smooth. Large, wrinkled brains are associated with thought—although no one actually claims to know why this should be, since, as we know from electronics, anything can be miniaturized. "Scientists tend to find what they expect," Miriam said. But she knew a University of Florida researcher, Robert Reep, who looked at manatee brains a different way: he notes that the percentage of the manatee's brain devoted to cerebral cortex—the portion that, in humans, is associated with thinking—is relatively high. It's comparable, in fact, to the brains of primates, and a markedly greater proportion of the manatee's brain is devoted to cerebral cortex than in animals like bats.

"To find all those lakes and creeks," Miriam said as we scanned for our study animals, "to know when the water level will go down and they will have to leave, manatees must be far more intelligent than people think." At Mamirauá, she said, the manatees are wary. "It's hard to catch them. The fishermen tell me, you can't blink, you can't breathe—they slink away." Miriam is convinced they have learned to avoid people, becoming increasingly nocturnal, just as some researchers believe beavers have done in America during this century, in response to hunters. No one would have expected that of manatees— but in Mamirauá, a lens into the Amazon itself, reality seldom conforms to expectation.

"The theories we have for many subjects are not completely true," Andrea added thoughtfully, continuing the conversation as we handed over the telemetry to Miriam. Last year, for instance, during the dry season, Andrea had witnessed an event undescribed in the scientific literature, which nobody would have believed.

Andrea, too, was a maverick. A sturdy woman of thirty with long black hair and pillowy lips, to the dismay of her family she saw forestry as her calling, and chose the fantastically complex mechanics of seed dispersal in the flooded forest as the subject of her master's thesis. The project required that she sit motionless beneath fruit trees for ten hours a day, ten to fifteen days at a stretch, observing who eats the ripe fruits as they fall into the water. One day, she was watching a group of brown capuchin monkeys at the side of a river at Jarauá. To her great surprise, she saw they were grabbing fish out of the water with their hands! Most primatologists believe New World monkeys don't eat fish, except for an occasional scavenged meal. No one had recorded monkeys actively fishing for them.

"That, I think, is why Mamirauá is special," Andrea continued earnestly. "Here we are always discovering things that we thought cannot happen—but here it *can* happen."

What were some other examples? I was surprised by Andrea's answer: "In the past," she said, "we thought that in a reserve you cannot have people living in it. But even we don't know that it's possible even now. The challenge is to discover this."

Here, where so many impossibilities come true, she and Miriam, Márcio and Ronis, are hoping for a new miracle: that today, people can rediscover how to live in a flooded forest without destroying it. And beyond that, they are hoping that these people will actively protect it.

As we traveled, scanning for botos and manatees, we passed their settlements. As at Tamshiyacu, the people, here called *caboclos,* come from mixed ancestry, with both Indian and Portuguese blood. They live simply. Most houses are wooden, with tin roofs, and only three rooms, lit by kerosene. Some families live on floating houses, but most homes are built on stilts. If the waters rise too high, the family simply raises the floor. Those who have cattle or pigs build for them floating pens called *marombas,* or else they bring the animals inside the house with them. Usually, the entire family—including, on average, five children—sleeps in one bedroom. The average family income here is $900 a year, which people use mainly to buy salt, sugar, cooking oil, powdered milk, and soap. But not a few now have televisions, powered sporadically by village generators or by solar panels, bringing them the news of slain princesses, of foreign wars, of fancy clothes and shiny new appliances.

These are the people traditional conservationists consider the problem. The first step in establishing a new reserve is usually to move the people out, at least into border or buffer zones. But again, Mamirauá defies convention. The reserve's founders hope instead that these people will serve as the forest's guardians.

Yet, despite federal laws prohibiting it, the people hunt manatees here. They hunt and eat endangered turtles. They fish in the reserve's waterways. They cut its timber. All of these activities are allowed under Mamirauá's management plan. For, the reasoning goes, if the people understand Mamirauá's riches are their own, they can protect this vast ocean of forest—its manatees, its dolphins, its fish, turtles, and timber—better than federal guards.

The Brazilian government had stationed two rangers to patrol Mamirauá's vast waterways—better coverage than in most Brazilian parks, Miriam told us, which average one ranger per 1,415 square miles. But the guards were powerless to stop the commercial fishing fleets that streamed in from as far away as Manaus and even Colombia. In a single season, outside fishermen had stripped one of the lakes of eighty tons of fish—and threw away all the catch but the tambaqui, wasting enough fish to have fed one-third of Mamirauá's residents for a year.

Enraged local residents, with the help of Mamirauá planners, created community patrols to evict the trespassers. In the first year of community surveillance, 1993, the amount of fish sold in Tefé taken from the reserve was cut in half. According to a 1996 report by the Overseas Development Administration, which helped fund the Project, "the invasion of lakes by fishermen from Manaus and Manacapuru, before very extensive, practically disappeared. The surveillance of these lakes, under the responsibility of the communities, has already proven to be effective."

Next, General Assemblies, in which representatives of all fifty-three of Mamirauá's settlements convened, developed a zoning scheme for Mamirauá's lakes. Some are assigned strict protection; others are reserved for subsistence fishing; and still others allow commercial fishing by community members. Logging is similarly restricted. No tree-cutting is allowed along the elongated levee banks known as *restingas*, which support the tallest trees. No logging is allowed except during the wet season, when the trees can be more easily taken out, lifted by the water, leaving fewer scars on the land. Lumbering areas are divided into sectors, where cutting is permitted only every thirty years. The people are working on an agroforestry program, similar to the Tamshiyacu-Tahuayo Community Reserve's.

But what the community agrees to do and what actually happens are not always the same. That was why we were traveling to Jarauá: Andrea needed to talk with the woodcutters there, to confirm that she had counted every log cut, catalogued every species. It is extremely important that researchers be able to accurately document what is really happening, Andrea stressed—and this demands that people trust her. "In 1993, when we were beginning the research, they were very afraid to talk with us," she explained. "They thought it was a vigilance." Rumors were rampant: some people suspected the researchers were secret police financed by the government to spy upon and punish them; others believed the scientists were conniving to sell all the fishes to Great Britain. Many feared, not unreasonably, that the researchers would use their data to outlaw all logging in the reserve. And many of Mamirauá's residents could not survive without it, which Andrea both understands and respects.

"There is a period of the year—May, June, July—that all the land is flooded so they can't plant," she explained. "Fishing is difficult because the fish are distributed into the forest. Logging is an important economic activity. So we are looking to find a way to let them continue this important activity."

The loggers cut with axes—chainsaws are only for the richest villagers. "The villagers dream with chainsaws," she said. To cut a tree with a chainsaw, and not an ax, would be, for most of the people here, an unimaginable luxury. But axes alone still kill forests. They felled New England's, after all. Here in Mamirauá, the large kapok tree was once highly sought-after for plywood, since it grows so big—"a huge tree, one of the biggest," Andrea told us. "But now, each year it's harder to find.

"When you talk with older people, they tell stories of big turtles and big pirarucu they used to see, and don't see anymore. They are worried about that, too, as we are. But it's not easy . . ."

The Project has attracted international interest and generous funding. New York's Wildlife Conservation Society funded the reserve's establishment with $4.3 million in 1980; it is now also supported by respected organizations including Conservation International; World Wide Fund for Nature; the Rainforest Alliance; Friends of the Earth. The universities of Oxford and Cambridge send some of their finest scientists here. Mamirauá has been proposed as a future biosphere reserve under UNESCO's Man and the Biosphere Program. And as the first sustainable development reserve in Brazil, Mamirauá provides the legal framework for the creation of similar reserves throughout the Amazon. "Other people are watching us," Miriam had told a visiting reporter for *Science* in 1994. "If it works, some of these techniques will be copied."

But still, many are skeptical. Michael Goulding spoke to the same *Science* reporter as did Miriam. He regretfully predicted that local people would eventually overfish and overlog, just as the outsiders did. For human greed does not always lust after oil and gold and diamonds; sometimes greed merely whines for a chainsaw. Sometimes it only wants a television, or a motor for the canoe, or a little extra meat to flavor the manioc. Greed urges good people to take just one more manatee out of season, or just a few more logs than the neighbors have harvested. And for us, with the produce of the globe in our markets and the splendor of the nations beamed to our computer screens, it seems monstrously arrogant to blame them.

"When we see a manatee, and think how long it spends to become an adult, we wonder how can they kill something so beautiful?" Andrea said. "But when you live here and spend days and days eating just one kind of food, you begin to understand.

"The people here, they like to live here. But in their minds, all the world is like this, you see."

We arrived at Jarauá the following day at noon. Jarauá is a little village of floating houses with potted gardens and stilt-legged shacks. In one of them, a green mealy parrot perched in a window and a satellite dish squatted on the roof. Gooselike horned screamers, who issue their un-earthly calls all day, hoisted their corpulent bodies from the water and flew ponderously to the tops of trees, gulping "Hoop! Hoop! Yoik-Yoik-Yoik!" A shimmer of parakeets twinkled across the sky, and the sun beat down like a hammer on a sheet of gold.

Only a half day remained until I would have to give the telemetry back to Miriam. We had seen many dolphins in the past two days, but none bore radio tags. Above a town called Pirarará, near a grassy area where Miriam often sees manatees, we had stopped to watch two botos fishing together, and also saw six tucuxis. Three botos had come to watch us as we'd stopped to tow an errant floating tower back to the middle of the river, and one had come to observe us delivering Ronis's refrigerator, spy-hopping at a distance of about thirty yards. At a cross-roads of rivers, where small fish skimmed the surface, we had seen five botos together. Most of the sightings, though, were single botos spotted in shallower waters; the tucuxis preferred the deeper channels. Travel-ing at the *Uakari*'s stately six miles per hour, we had counted what we thought may have been thirty different botos since we'd left Boca. In the silence of the telemetry, though, I read a lesson: in the Amazon, you never get the answer you expect.

Andrea and Dianne took the speedboat downriver to talk with vil-lagers. Miriam and I debarked to visit the turtle researcher Augusto Teran in his dockside lab. A small, passionate Peruvian, he was study-ing the freshwater turtles here for his doctoral thesis. Miriam translated as he told us in Portuguese that he had captured and marked fifty-two turtles the night before in his nets. He sets the nets across the channel, at the deepest point, so they don't catch caimans. For the turtles' safety, he checks his nets every four hours—the reptiles can go without oxygen for six hours, but the main danger is piranhas. Enthusiastically, his dark eyes aglow, he showed us the shells of the six species found here: the largest are the three *Podocnemis* species, who tuck the neck sideways beneath the shell, like the one Gary and I had bought from Don Jorge.

The biggest shell was, for a freshwater turtle, enormous: the largest species, *Podocnemis expansa*, known here as tartaruga da Amazônia, can grow a shell thirty-five inches long. They eat leaves, fruit, and seeds, hunt fish, shrimp, and crabs, and scavenge dead fish, Augusto explained. Augusto showed us the shells of the two smaller species, *Podocnemis unifilis*, known locally as tracajá, and *Podocnemis sextuberculata*, or iaçá. There are also three other kinds of turtles: the land tortoise known as jabuti, the freshwater perema, and the weird, side-necked matamatá, which hunts underwater motionless with open jaws, waving a fleshy protuberance on the tongue to lure fish. But Augusto's primary focus is the *Podocnemis* species.

He explained his work: Once Augusto captures a new turtle, he measures it, bores a hole in the back of the shell, and affixes a small yellow plastic tag with a number. The tiniest turtles are too small for tags, so he inserts a minute electronic marker under the tail. From this research he has been able to determine the rate at which they grow: one he caught in January then weighed 10.5 ounces, and by September it had grown to weigh 15.5 ounces. When locals catch his tagged turtles, they bring them to him so he can record their growth; and then, of course, he must let the fishermen take the turtles back home, where they will be cooked and eaten. Otherwise, the people would never bring them in.

This had been the fate of all the animals whose empty shells we now beheld—a fate that now threatened their existence. The biggest *Podocnemis*, the tartaruga da Amazônia, has been nearly hunted to extinction in Mamirauá, Augusto told me. "So the people turned to the next-largest species," he said, the tracajá. Now that species has been decimated to the point that people are hunting the little iaçá—and now they are mainly catching only the juveniles. "The people are wild for turtle meat," he said.

The people set their nets for turtles along their migratory routes. The turtles behave like manatees, Miriam translated, and leave the lakes when the water drops; *Podocnemis* males prefer the shallow waters, where nets are easiest to set, so mostly males are caught in this way. But the females are even more easily captured on the sandbars of their nesting beaches. And their eggs are taken, too: of seven nests Augusto recently found, two had been destroyed by lizards and five by people.

Even worse, the eggs that would normally hatch into females are especially vulnerable. Like many other species of reptiles, the sex of an individual is determined by the temperature at which its egg incubates; the hotter eggs hatch into females. Because these eggs are easier to dig up, Augusto said that in his counts, male turtles now outnumber females six to one.

As we were talking, a fisherman came into the office, carrying a large, thrashing tracajá. Augusto's eyes lit up. He measured the animal's dark shell at 17.94 inches, and it weighed in at 30.8 pounds—the largest of the species he had ever seen, a female.

How old was she to have grown to this size? She could have been thirty-five or forty, Augusto answered—about my age, I realized. Augusto laid her on her back, her gray eyes staring upside down, and he spread her scaly back legs apart to feel the soft area at the inside of her thighs. ("Poor thing!" said Miriam—who hadn't evinced sympathy for this turtle's fate in the stewpot, but pitied her gynecological exam.) Were there eggs inside her? Tracajás can lay more than forty-five eggs at a time, Augusto said, and produce up to three clutches a season. But she was empty now.

I imagined how she might have heaved her body from the water onto the sandbar on some recent, full-moon night: driven by millennia of turtle-knowledge, she would have crawled perhaps a quarter mile along the sand before selecting the right spot. Then she would have begun to dig. Her strong, horny rear legs and her curved, scaly feet would have sent sand flying, its grains glistening like dew in the moonlight, until the hole was two feet deep. As her gray eyes flowed with tears—the turtle's way of protecting the cornea from flying sand—she would have birthed dozens of perfect, leathery round eggs into that cool hole, and then, again with her back feet, covered them with sand; and finally, her momentous job complete, she would have dragged her tile-smooth plastron back over the beach, the toenails of her back feet carving *S*'s into the sand. Finally, the water would have risen to reclaim her body, weightless and free—only to be caught in someone's net, and to lie here upside down, awaiting her fate with turtle-patience.

Augusto was now speaking earnestly to Miriam in Portuguese. I guessed correctly what he was saying: he wanted to buy this turtle from the fisherman to set her free. Of course, he couldn't start buying turtles from people, or soon he would set up an industry—he knew that as well as Miriam did. But nonetheless, Augusto longed for her life.

I did, too. Mamirauá doesn't get enough visitors to make selling turtles to tourists an industry. Perhaps, I suggested, I could buy the turtle for him?

"It's only one individual," Miriam said. "It wouldn't make any difference—it wouldn't do any good."

"It would do a lot of good for this one," I answered. "But I don't mean to undermine the Project, or the management plan, or to do anything that could endanger other turtles. Could you ask Augusto if he thinks this would work?"

Miriam translated. "He thinks it's a good idea," she said.

"And you, Miriam? What do you think?"

"Maybe," she said thoughtfully. "Maybe it would set a good conservation example, for the people to see a turtle set free."

I asked her to find out how much it would cost.

"He thinks it's worth fifty reais." Roughly fifty dollars.

I was stunned. Gary and I had bought the *Podocnemis* from Don Jorge for well under the equivalent of $5. Were the fishermen ripping me off because I was American? No, said Miriam; the people here so prize turtle meat that Augusto had once seen a man trade a turtle for a propane tank.

I had 74 reais left, plus a $100 bill. If I bought the turtle, I would have to borrow money later from Dianne to pay for the rest of our stay in Brazil—including the petrol for our speedboats here, taxi fare for the airports at Tefé and Manaus, departure tax, and anything I ate from the day we left Tefé to the day I returned to the States. Buying any souvenirs from this trip would be out of the question, and I would arrive home in worse debt than I already was.

We waited for Dianne and Andrea to get back from the village.

"I'm spending my last big bill on a turtle," I announced when the speedboat returned, "if you can loan me money for gas."

Dianne did not approve. "It's trade in wildlife, and I don't like it," she said sharply. That was an understatement. Dianne had personally witnessed the suffering caused by the burgeoning trade in wild animals. She had once been summoned to Thailand, and later to Borneo, by the International Primate Protection League to care for six baby orangutans confiscated from an illegal animal dealer; two of them had died in her arms. With Shirley McGreal, the director of the League, Dianne had bravely testified in court, despite phoned death threats, to land that dealer, Matthew Block, in jail. Dianne did not want to see this rare turtle eaten any more than I did; but trade in wildlife was, to her, an intrinsic evil, far worse. She wanted no part in stimulating further trade.

But would it? Or would my purchasing the turtle have some other ill effect that I, an outsider, could not foresee? Surely the aid workers who taught *caboclos* to grow cattle and corn all over the Amazon believed they were doing a great service, as did the Christians who robbed the Indians of their gods and their languages. In the Amazon, it seemed that nothing—the biological world, the political world, the very laws of the universe—operated under the rules I had learned in the States.

I wanted to know what Andrea thought. Not only was she one of our hosts, she was also the expert on local resource use. We decided to discuss it onboard the *Uakari*, leaving the gray-eyed turtle on her back in Augusto's laboratory. But before we left, Augusto tied her to a table leg with a string threaded through the hole in the back of her shell. "She's worth fifty dollars," he said in Portuguese, "and I can't have her escape now!"

Andrea felt we should not interfere. "We should let them eat it," she said solemnly. "It's a different way to conservation. They can use this resource according to their needs." Mamirauá's Management Plan, she reiterated, states clearly that local people should be able to continue to hunt food animals in the reserve—even though it allows killing of animals whose slaughter is outlawed by federal decree.

"But what about setting a conservation example?" I asked.

"They will only think we are funny," she said.

Our discussion continued for perhaps half an hour. We spoke in English, so the fishermen would not know what we were saying; we did not want to give them the idea that a lucrative market might exist selling these turtles to foreigners. But Augusto, even though he speaks no English, was well aware of the turn the conversation was taking as he read our faces and voices. I could see his distress as the fire in his eyes began to die. Years ago, I had read an article by a prominent conservation biologist with New York's Wildlife Conservation Society, which today funds much of the work at Mamirauá. "Individuals," the author wrote, "cannot mean that much when you have to do large-scale manipulations of populations, as a conservationist sometimes must." But this turtle's individual life was far from meaningless: in deciding her fate, we grappled with the promise and the plight of the Amazon.

"Once I had this happen with a manatee," Miriam told us. "It had been harpooned. It would have survived fine if we had released it. I had to watch them kill it in front of me."

My father was an Army general; and I knew, with heaviness in his heart, he would have made the same decision—to sacrifice the individual to the hope of a greater good.

"The management fact is, this animal has already reproduced," said Andrea.

"But doesn't she have value as a study animal? As the largest Augusto has ever seen?"

"Not really."

"But she could lay eggs again—maybe even this season," I said desperately.

"Maybe she's even past reproduction age," said Miriam.

"A menopausal turtle?" I asked. Andrea and Miriam thought this was hilarious, and deemed it worth translating for Augusto. But he didn't laugh. Instead, he mumbled something in Portuguese, as if talking to himself. I asked for the translation. "He says maybe she can still teach the others," Miriam relayed.

But the issue had already been decided. I felt utterly defeated. Miriam had been right: this was just one turtle. In the forty-five minutes that I had spent worrying about her fate, if the scientists' calculations are correct, a tract of Brazilian rain forest the size of 315 football fields had been destroyed—a rate of 5 million acres a year. Most of it would go for particle board to customers in the United States, Europe, and Japan, and most of the profit would benefit executives in Malaysia, Indonesia, China, South Korea, and Singapore. What I had set out to do—to save one turtle from the pot—would do nothing to quell these huge foreign appetites. But perhaps Mamirauá's management plan might.

Every one of us—Miriam, Andrea, Dianne, Augusto, and I—wanted the same thing: to save this toweringly cruel and nourishing dawn world from fading to twilight. "In my opinion, you can preserve biodiversity only if the people want it," Márcio Ayres once said. "If the president says, 'I'm going to preserve,' that won't mean much if the government changes in four years. But conservation by the wish of the local people cannot be changed. Then it is a political movement among the local people, a way to get their rights. If the movement comes from the people, it will be very hard to stop."

Later, back at the floating house, we told Peter about our discussion about the turtle. "That's why I work on this project and not others," Peter told us. "The correct premise to start conservation projects is that people are important. I wholeheartedly embrace Mamirauá, as I think this is a conservation project of importance on a world scale, of global importance. I'd like to think of my children and grandchildren knowing that I was a part of this thing. I'd like to think that Project Mamirauá will be that successful."

The issue was decided. Dianne saw that further discussion was futile. "She's menopausal!" she said. "Whack her, I say! I'm hungry." And then she let loose her pirate's laugh and lit another Newport.

The following night, Dianne and I sat in the anteroom of Ronis and Barbara's floating house, eyes glued to the television. It was playing a Japanese video of Ronis making a caiman vomit.

On the screen, with a PVC pipe lodged between the caiman's jaws, Ronis inserts a tube into the reptile's stomach to flush it with water and dislodge what it has been eating. Its favorite food seemed to be the same as the dolphins': a member of the fish family Loricaridae, a bottom-feeder. The caiman's jaws are taped shut for the procedure, "to restore more calm," Ronis explained to us as we watched. But the caiman on the screen is not calm. The five-foot reptile suddenly leaps off the table, dislodging the pipe, and thrashes on the floor, snapping. "Some problems," Ronis said matter-of-factly, again narrating the film to us. On-screen, Ronis picks up the thrashing reptile firmly but gently. His greatest fear about these powerful, prehistoric animals seems to be that he will unduly alarm them; this strong, outwardly macho man seems humbled and honored that he is a traveler in their universe. On this night, after watching the film, we visited that universe with him.

We embarked in Ronis's motorized canoe. It is painted green, he explained, because white alarms the caimans, as bright light alarms the fish. On our way to the caiman lake, we turned off our headlamps, for we had already discovered this on the boat ride from our floating house to Ronis's. Fish had leapt out of the water thick as bow spray— an upside-down waterfall of fish, spewing skyward as if leaping for the moon. For this reason, Ronis keeps an array of goggles at the house; people are not uncommonly injured when leaping fish smack them in the eye. "The first time I was here," Ronis said, "it was impossible to see the caiman because so many fish!"

We entered the lake. "In this lake, wet season, can see twenty-six hundred caiman!" Ronis told us. Mamirauá boasts thirty-nine caimans per square mile—the highest density in the Amazon. Sixty percent of them are the largest species, the black. Around us, a great wall of red eyes stared into the darkness, glowing balls of blood, a meniscus of eyes. In our spotlights, we saw the great armored heads. They seemed immobile, waiting with an elegance no mammal knows.

They have seen so much time. They each own a part of an ancient knowing; a timeless fire glows in their eyes, patient as a volcano—and when necessary, as sudden. The spectacled caimans sink from sight like submarines, but the black caimans vanish in a flash of water-shudder, as if they somehow call the water up to surround them rather than sink beneath it, the way a magician might disappear into a puff of smoke.

Ronis uses a 400,000-candlepower searchlight, powerful enough to show us that both the shores and the shallows are covered with caimans. "With common species on land, and black in water, the area can support great density," he said. Next month, they will be nesting. "It's a marvelous species," he said, "eight million years old."

Only eight million? Ronis and I had hit a numbers glitch before. He had told me the mother caiman stays with her young eight years, and I had been astonished. I had known that some crocodilians care for their young, but eight years! The figure rivaled orangutans, who nurse their babies for that long. When I mentioned this to Peter, he assured me Ronis must have meant eight months—still impressive. Once, he told me, Ronis had asked to borrow his speedboat for a year. He had meant a day. I asked Ronis to write the age of the caiman tribe, and he wrote, "80,000,000." Eighty million.

At that moment, a four-inch fish with silvery eyes flipped into the boat and flopped around our legs. I gently pinned it with the toe of my shoe. "Poisonous spine," said Ronis, lifting it gently by a fin, revealing the needlelike projection from the pectoral. "Can be very dangerous." He tossed it overboard. Later, a dogfish with inch-long daggers for teeth flew into the boat. In the morning, my sneakers would be covered with the silvery scales of other fishes, whose entry into our craft we had failed to even notice.

Ronis was eager to catch a black caiman, his favorite of the three species here. He had his eye on several larger animals, over ten feet long. "This species so calm," he said. "The other, no. This lucky, because this is bigger." He lunged with his loop, but the caiman winked away into the water. Lightning flashed, the sky throbbing white-hot, answering the reptiles' eyes. Fish leapt to join the lightning. The world was remaking itself, its plan a complete mystery.

We slipped along the margin of the lake to watch the spectacled caimans slither into the forest. They are more graceful on land than I thought possible, moving snakelike until they are swallowed by the grass. The spectacled are so called because their eyes are raised, Ronis explained, while the blacks are named for the four black spots on the lower jaw. And with this, he looped the noose around the neck of a four-foot black caiman and hauled it squirming from the water. Its long tail flapped vigorously until Ronis held it against his pole and muzzled the jaws shut with masking tape. He held the creature out to us to let us touch it. Its belly felt like the tile on the bottom of a swimming pool—cool, clean, impenetrable, permanent—like the look in the caiman's eyes.

What goes on inside a caiman's head? Earlier, I had spoken of this sort of thing with a friend, David Carroll, who has made a life studying turtles in New Hampshire. "You can't speculate what goes on in their brains too much," he had said to me as he released a five-inch wood turtle back into the alder thicket where he'd found it. Humans, we agreed, are unduly impressed with the fact that we think; but animals *know*. "But they have such a history, and they're united to it in a way we are not. Whatever he knows," David had said of the turtle, "goes back to two hundred million years ago, to the first turtle. A lot of his messages, I think, are from that reserve, and beyond."

Ronis caught many more caimans that night. We didn't count them, for we had left time and sequence behind. Each moment was a fresh wonder, new, from the timeless water-womb of all beginnings. We felt like space travelers to another universe that night, with the bowl of the stars reflected in the water below, and the caimans' eyes glowing on the horizon like a thousand red suns. I thought of the verse from Psalm 8: "When I look at Thy heavens, the work of Thy fingers, the moon and the stars which Thou hast established, what is man that Thou art mindful of him, and the son of man that Thou dost care for him?" And I remembered something Andrea had said the night before we visited Jarauá. We had watched the darkness rise on the river, as if the night itself were rising magically out of the Encante. "Everything here is big, and we feel like a mosquito," Andrea had said. "In town, we feel we are so important—and here, we realize we are nothing."

Dianne and I both held a lovely two-year-old black caiman Ronis captured from the bank. Its jaws did not need to be taped shut; the creature rested placidly in our hands. Our flashlights had caused the caiman's eyes to glow red, reflecting from the tapetum lucidum, a light-gathering mirror in the eyes; but actually, we now saw, the animal's eyes are golden, full of the light from a thousand stars. I released the tiny caiman into heaven's waters, and asked it silently to carry my blessings with it to the Encante, back to the beginning of time.

We never did get a signal from the telemetry. Miriam assured us this was no fault of ours; she often searched for days without finding her manatees, and Vera had, after all, just spent eight days the month before searching unsuccessfully. Possibly the transmitters on the last three botos had now gone dead, too.

After we returned the equipment, we continued to watch the dolphins every day. Our backs and legs ached from the boat's hard, backless benches; our skin blistered; we stank with sweat—and all of this vanished with the sight of a fin, a face. Every day they came, surrounding us, six, eight, ten or more at a time. They seemed to recognize us, though we didn't recognize them; if Ruffles or Shika or Scar ever reappeared, the water and glare kept their identities secret.

Each day brought some fresh wonder. One morning, tucuxis joined the botos, and one of the small gray dolphins rolled and rolled at the surface, over and over like a child twirling herself dizzy. Another morning, the whole river seemed to be gathering its breath for song: the tucuxis splashed, the dolphins blew, and then from the forest, the howler monkeys began to chorus, their voices growing in intensity like a gathering wind. Their song rose and sank and pounded, its melody curved like the botos' smooth bodies. Then their song died, and almost instantly, the dolphins vanished, taking even the waves on the water with them.

Each surfacing brought a physical thrill, like the sudden drawing in of breath, or the rush of a shudder. With each sighting, the waters opened with new promise. Yet at the end of each day our longing was undiminished. And when we left Mamiraná, we knew our journey was not over.

What did I want from them? No longer did I hunger to map their travels, Point A to Point B. No longer did I need to count them, to time their surfacings, to measure their breaths. I no longer wanted to pursue them; now I wanted to join them. I longed to swim, like them, with my nostrils above water and my lips below. I longed to look into a face, and know that face. I wanted, in short, to partake of their grace—not simply the grace of their effortless beauty, but the grace of a benediction. But what form this might take, how or where I might seek it, I had no idea—only that this still awaited me, an unspoken promise.

Later, back in New Hampshire, I remembered Dianne's dream of lovers and diamonds after Ricardo first consulted the dolphin spirits, back in Peru. "Glass is good luck for you," he had told her. "So is clear water."

Perhaps the dream had offered that promise.

The Moon's Tears

"THE SUN AND THE MOON WERE ONCE LOVERS," THE YOUNG man tells me, "but they were separated by the god Tupa." We are standing waist-deep in the clear blue waters of the Tapajós, watching for dolphins. "They were impossible lovers," he says. His dark eyes dart into mine to make sure I understand what he means.

"But sometimes they meet. Not often. It is called . . ." He is looking for the word in English.

"The eclipse," says Dianne.

"Yes. Eclipse," he says. "But it is rare that they meet, very rare."

Exiled to a realm separate from her beloved, the moon wept. And according to the creation myth of the Mawé Indians of Amazônia, that is how their world was formed: "They say it is the moon's tears that formed the Amazon," Felicio says. "It is the river of impossible love."

BURNING

The Apostle Paul wrote to the Corinthians about longing for the Lord: "For now we see in a mirror darkly. . . ." He might as well have been writing of the dark mirror of the Amazon, concealing all but fleeting glimpses of the dolphins. Its mirroring qualities are both the danger and the revelation of the water-world: we are blinded to its depths by the image of our own reflection. No wonder the authors of the Bible painted God's appearance as a man, only omnipotent. No wonder river people believe the Encante is a world like this one, only brighter. We see our own desires reflected and magnified in the river's polished surface—but then we miss the truth of its depths.

To penetrate the surface, to pass through that dark mirror, requires a soul's rebirth. Paul hints at this promise as he writes of the future: ". . . but then, face-to-face." By what baptism could I come face-to-face with the botos? The answer was shockingly simple: by meeting them in clear water. Clear water: that was the message in Dianne's dream.

I had found no published studies of dolphins in the Amazon's clear-water tributaries. But we had seen film clips of botos in clear water, and discovered they had all been filmed in one location: beside a stretch of white sand beach along the Rio Arapiuns, in the Brazilian state of Pará. Crews had always filmed in the dry season, since the water was shallowest then. Now, in November, the water was nearing its lowest point.

In our e-mail correspondence, Vera provided us a contact: in the little resort village of Alter do Chão, an American expat, David Richardson, knew where to find the dolphins, and was said to have sometimes swum with them in the clear waters of the Rio Tapajós. Both Alter do Chão and the site on the Arapiuns were within a day's travel from Santarém, the third-largest city in the Amazon, two degrees south of the equator.

But to get there, we would have to fly through skies full of fire.

With the hot, dry weather brought on by El Niño, the fires set each dry season by loggers and ranchers throughout Amazonia were burning out of control. The *New York Times* reported that at least 10 percent of the 2 million square miles of Amazon forest had been destroyed by fire. "Raging fires in Brazil dwarf the ones in Indonesia and Malaysia and could ultimately pose a larger threat to the rain forest's rare and uncharted animal species, the supply of breathable air, and even the world's climate," we read in an ABC News story on the Internet.

Forest fires in Indonesia and Malaysia had been burning since mid-July, to international dismay. Smoke from a 4.2-million-acre blaze completely obscured the island of Java from satellite photos; in September, the smoke had been so thick that an airplane crashed and a supertanker collided with a cargo ship on the same day. Those incidents made the Asian fires headline news for weeks. But the Amazon fires got comparatively little press.

What we *had* heard, though, was disturbing. Vera e-mailed us that the low water again meant hydropower was rationed. Phone service was spotty; we never could reach David Richardson. Manaus, we had heard, was engulfed in smoke. Another friend forwarded correspondence from a colleague in Brasília, who said some days the smoke was so bad that planes couldn't land; many people were hospitalized with breathing problems. I bought an inhaler for the trip.

Tapajós and Arapiuns Rivers

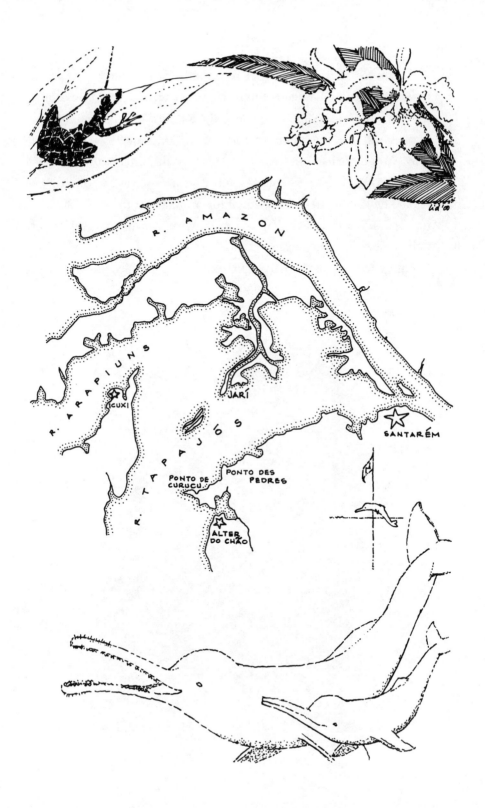

R. AMAZON

R. ARAPIUNS

ICUXI

JARÍ

R. TAPAJÓS

SANTARÉM

PONTO DE
CURUCU

PONTO DES
PEDRES

ALTER
DO CHÃO

But most Brazilians didn't seem to think the fires were unusual. At the Miami airport, I asked the Brazilian ticket agent at the Varig counter for news of the Amazon fires; surely the airlines would know the state of the skies. "Oh, the fires burn, the rain puts them out," she said cheerfully. "You come back and show your tan!" At the airport at Manaus, we met a German, the president of a timber firm, who was flying to Belém; he said he had heard nothing out of the ordinary. At the Hotel Monaco, no one could furnish us any information; we began to wonder if the few news reports had been exaggerated.

But when we arrived in Santarém, we could see the land was

parched. Charred trees and singed fence posts lined many of the roads; in some places, even the red soil was blackened by fire. "It's what they do this period of the year," said the cabdriver who took us to our hotel. "Just the fields are supposed to be burnt, but the fire always escapes. It's normal. Only this year, it's too dry."

Because the river was so low, electricity was rationed to twelve hours a day—which twelve hours varied daily. Because our room at the Hotel Tropical was *aircondicionado*, it had no screens; when we opened the windows at night, mosquitoes and moths flooded inside, and I wondered whether bats might soon follow to hunt them. But our winged roommates were the least of our problems. We still couldn't reach David Richardson. When we dialed his number, we kept getting a recording we couldn't understand. Overhearing our distress, a guest at the hotel, a wholesome Belgian woman with short blond hair, decoded the message: his was a cell phone, and it was now out of range.

Isabelle Druant, we learned, was a forty-year-old agronomy student who had spent a year teaching English in Brazil at Belo Horizonte. She had come here on vacation. So far, it wasn't very relaxing. She had made arrangements to travel on a boat named the *Sim Blanco*, but then discovered that it had sunk ten years ago. When it had entered the harbor at Manaus, all the passengers had rushed to one side to see the beautiful city, and the boat flipped over and sank. Isabelle then booked herself on the *João Pessoa*, a sturdy-sounding, 345-ton, three-tier passenger and cargo vessel. Normally, it could accommodate 425 passengers, but now it carried only 150 because the entire bottom tier was jammed with cargo: coffee, wheat, farina. The boat was so overloaded, she said, that when she first boarded, she had climbed up a plank, and once everything was aboard, the angle of the plank had reversed.

When they reached Santarém, she and the other passengers had planned on spending their first night on the boat, which was included

View out the bus window: the Amazon burning.

in the fare. But that afternoon, they were advised to leave the vessel, although they were assured they could leave their luggage aboard. Isabelle thought the boat was listing strangely, so she removed her luggage from the cabin.

When she came back, she found the boat had sunk—along with three cars, two motorcycles, two horses (who swam ashore), the coffee, wheat, farina, and all the passengers' luggage—including all the worldly possessions of at least one family who was moving to Santarém. We later encountered the captain at a restaurant, who told us the river was so low the boat had hit a dock piling, breaking the bilge pump. There was no insurance.

Isabelle relayed this story while Dianne and I wolfed down sandwiches on the Hotel Tropical's poolside patio—I had ordered *vegetariano* and had been issued a ham-and-cheese. We told her about our project. We needed a translator. Would she like to join us? Isabelle said she'd be delighted.

I asked her for news about the Amazon fires. "Fires? I haven't heard anything about fires," she said.

The next day, the three of us took the bus to Alter do Chão, thirty-one miles away, to look for David Richardson. A taxi driver had told us we could find him at the Indian museum there, the village's sole cultural attraction.

The museum is a big yellow cement structure with a red tile roof, a block from the bus stop along the red dirt road. We were met on the steps by the cleaning staff. David wasn't there, they told us, but his wife, Maria Antonia, was. A fine-boned woman with horn-rimmed gold glasses and shoulder-length black hair, she told us David was in Manaus till the end of December. She knew nothing about the botos. Pressed, she finally gave us a cell-phone number for Manaus. She didn't offer us use of the museum phone.

We found a telephone outpost behind a pile of rubble next to the police station. David's cell phone was turned off. But thanks to Isabelle, we discovered that the pretty *telefonista* knew David's two boatmen: one was named Braulio and the other Simão. They live in town, she said. Everyone knows them.

We walked downhill along the dirt road to a little open-air restaurant near the big cement church that presides over the village square. Dianne was hungry. While we shared fried fish, isca de pirarucú, we looked out over striped beach umbrellas and thatched cabanas gathered around the turquoise Lago Verde, feeding into the blue Tapajós. Great gnarled mango and cashew trees, laden with fruit, grew parklike in the white and lavender sand.

The proprietor of the restaurant directed us to Braulio's house. He lived two doors away. Isabelle stood in the courtyard and clapped her hands: *"Oi da casa!"* she shouted, the local equivalent of ringing a doorbell that isn't there. A broad-faced man with gold teeth, thick glasses, and cropped gray hair leapt out of a hammock and limped to greet us. I asked Isabelle to explain our project, and handed Braulio my card. He stared at it intently. It was upside down.

Could he help us find the botos? You can sometimes see them right here, he answered, but they show themselves briefly and then hide. The best place to see them, he said, was a place called Ponto de Curucu, a spit of white sand named after one of the frogs you can hear calling at night. Did Braulio think the dolphins would let us swim with them? Nobody ever tried except for David, he replied.

So together we walked across the sandy bottom of the shallow lagoon to his fifty-eight-foot-long wooden boat, ambitiously named *Gigante do Tapajós,* and sailed away.

After we left Alter do Chão, the sandy banks of the Tapajós were deserted. We saw no cement houses, no stilt houses, no floating houses, no other boats—just blue water, white sand, and then forest. Within an hour, we were there. Braulio carried the anchor to the white sand beach and we sat beside it in a row, like spectators at a sporting event, waiting for the botos to appear to us on cue.

This is impossible, I thought.

And then they came.

There were two of them. Both were charcoal gray. We had read that botos tend to be darker in clear water, the pigment perhaps protecting the skin from the sun—and with their bulging melons, they looked like dapper English businessmen dressed in gray flannel and bowler hats. One began to spin in the water, perhaps thirty yards away. His flippers, surprisingly, were pink. As he twirled on his back, he opened them, as if inviting us to join him.

Dianne and I stripped off our shirts and slacks to join the dolphins in the water. "Are you sure there are no piranha?" asked Isabelle. Of course, I had no idea; but David Richardson had swum here safely. "No piranha!" I announced confidently—and, reckless, rushed into the clear blue water.

Dianne and Isabelle joined me. Isabelle hoped Braulio wouldn't look. "Just three white chicks standing in the Amazon in our underwear," I said. "What is so unusual about that?"

The botos came within ten meters of us, clearly curious, but always keeping to the deeper waters beyond. Never had we seen a clearer view: for the first time, we could see the botos' flippers beneath the water, beating like wings.

For an hour, they swam back and forth in front of us. We saw no unusual behaviors, nothing we hadn't previously recorded, but now we could see them—the flippers, the head and the back and the tail—*all at once*. Their bodies were perfect. For the first time, they were not fragments, but whole. We stood in the water, transfixed.

After they left, we sat on the beach, limp from excitement. Only then did we notice the fire. A change in the wind brought its big cinders floating toward us like a flock of burnt butterflies. On the southeast horizon, we could see a column of smoke rising, fat and malevolent as a tornado, bruising the afternoon sky purple. The ashes grew larger and took on more shapes, like ghosts of the creatures the fires had displaced: sparrows, caterpillars, spiders, leaves, frogs. We asked Braulio about the fire. "It starts for farms, then burns out of control," he told us matter-of-factly, through Isabelle. Then he announced that night was coming, and we should head back. We left the smell of smoke behind.

It was dusk by the time we returned. The power was out, and everything was dark, quiet, and warm. As we walked back over the beach, the little frogs in the green lagoon were singing. "Curucu! Curucu!" they cried, as if affirming this was the place to which we had been called.

We made arrangements to move our base from Santarém. Isabelle picked our new hotel: Pousada Alter do Chão. It faces Lago Verde and the naked red mountain that gives this town its name. "Alter do Chão" means "above the floor" in ancient Portuguese, we were told. The hotel has a wonderful restaurant on its porch. Roses bloom in the doorway and bougainvillea drapes from the roof. A Telepará pay phone, which we never were able to work, yawns its yellow mouth at the lake. Realizing the fan would be of little use, for now electricity was rationed to only eight hours a day (which turned out to be merely a theoretical maximum), we chose what we hoped would be the coolest room, fronting on the porch. We imagined that if we took cold showers and then lay naked and motionless on our beds, we might not swelter. In this we were wrong. But our choice was a good one. Our *pousada* was lovely and tranquil. The town hosts only about six hundred full-time residents, we were told, and in the mornings we woke to the sounds of whisking brooms and crowing roosters. In the evenings, we listened to the night songs of the frogs: some clatter like glass beads in a jar, others thrum like a drum skin rubbed with a wet finger, and still others, our favorite, repeat in winking peeps the name of the place we meet the dolphins: "Curucu, curucu."

Our motherly, dark-haired hostess, Socorro, which means "help," was aptly named. In back of her open-air kitchen, where she prepared meals on two large gas stoves beneath a tin roof, she had filled a small courtyard with orchids. There were more than two hundred of them, and she knew every individual intimately. Sometimes we would accompany her on her rounds as she watered them, and she would point out to us the unique graces of each: *"Como bailarina,"* she would say, showing us leaves that cascaded like a lacy waterfall; *"Pecino,"* she would note, pointing out blooms as tiny as the head on a dressmaker's pin.

Each of them, she had rescued from the fires. "One day, three years ago, I was out walking in the jungle near the hotel," she explained to us through Isabelle, "and I saw all these beautiful orchids in the trees. But

Socorro D'antona Machado with orchids
she rescued from the fires. (Photograph by Sy Montgomery)

the next time I went to see them, they were gone. The dry-season burns had killed them all. I wanted to cry." So the next year, and each year thereafter, she has made the same pilgrimage into the forest, to rescue the orchids before the fires start. She removes them from their perches—they are epiphytes, perching harmlessly on boughs—and notes carefully where she found each one. She takes them home and transfers them to the hollowed-out halves of coconut shells. She feeds them with the burnt seeds of the açaí palms that grow in the front yard.

A casualty of the fires: a dead tamandua, an Amazon anteater.

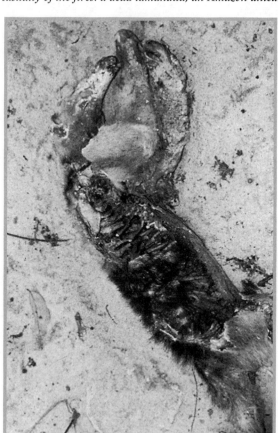

And there they stay, until the jungle grows green again. Then she takes them back to where she found them. "Three years ago, there were far fewer fires," she said. And now the river was the lowest she could remember: "This year is the worst dry period. I've seen pictures. There has not been so much beach since 1953."

She worried about us the way she worried about her orchids. She was excited to learn about our project—she had often seen the botos in Lago Verde, and thought them very beautiful. But she was a bit concerned as well. She teased us—the botos, she said, might want to take us away—especially Dianne, who, despite the heat, always appeared at breakfast as coiffed and fresh and beautiful as her neatly folded clothes. "Careful of those dolphins," Socorro said to Dianne with a wink. *"Cuidado com o boto."*

We returned to Santarém to stay one more night, for the next morning, a Friday, was the day of the regional market at the Mercado 2000. Vendors of fish and produce come to the shopping center from all over the region, and I was eager to talk with the fishermen, to hear if they had stories about botos. Isabelle was particularly interested in learning which medicinal plants were for sale.

We spoke with a well-muscled, gray-haired vendor standing by a tiny stall whose wooden shelves were crammed with bark, leaves, birds' nests, seeds, and bottles. In Portuguese, he introduced us to some of his inventory. These sesame seeds yield a fluid to prevent stroke, he explained. This bark, invirataya, can be burned to evict evil spirits. This pod, jucá, would cure intestinal ills. The grassy bird's nest, which he called eva de chombo, would yield an infusion to strengthen the liver.

On and on he explained his elixirs: a tea for gastritis, a poultice for warts, syrups for coughs, baths for infertility and infidelity. All of them Isabelle recorded with great care. Finally, I asked if he had anything

A vendor selling traditional medicines at the Santarém market.

that would help attract botos. He looked confused. He reached into a paper sack behind the counter and handed me the dried head of a viper. "This attracts botos?" I asked through Isabelle. No, he had misunderstood; the snake head, he explained, attracts money.

Isabelle and I tried to clarify, but my request was so odd that he had trouble believing that attracting a boto was really what I sought. Next he brought out a shriveled brown object, which looked like a six-inch piece of chewed rawhide. "What is it?" I asked. "It is twenty-five dollars," he answered. It was the dried vagina of a boto, he explained, and from this, a *curandeiro* could make a perfume that would help me at-

tract a boyfriend. Horrors! "Tell him I have a man, I have a man!" I implored Isabella earnestly. If I had a man, and didn't want money, what could be my problem? He considered for a moment. "Ah," he said. A light came into his eyes that showed that help was at hand: He brought out a four-inch piece of what looked like ligament. It was a boto's penis. "Infuse it into wine or liquor. The one who is weak must drink it," he said. "You will have great sex!"

I thought it was time we adjourn to the fish market.

We spoke with a handsome young fisherman of African ancestry as he sponged off the white tile counter, preparing to close shop for the day. We spoke of the fish at first, and the dry season, and finally I asked about botos.

"When the boto comes out of the water, he dresses in white and wears a hat," he told Isabelle in Portuguese. "When he takes off his hat, he becomes a dolphin again." He said he knew this was true because he knows someone who had a baby by a boto: the child is so white he cannot go out in the sun.

His companion, the owner of the stall, chimed in. In addition to the white clothes, he told us, the boto also wears shoes of kari—armored catfish. And the female boto calls people into the water, especially men who flirt with them. This doesn't happen here in Santarém—only in the country, he said. But his own father saw: as he was coming back from fishing, when they lived at a place called Lago Grande, he saw a man dressed completely in white, who jumped into the water and turned into a boto before his eyes. And yes, he assured us, women become pregnant by botos, but their babies have holes in the head and die very quickly.

It happened to one of his cousins: a boto had come to her one night, disguised as her husband. But her husband was away fishing. They made love as if in a dream. Nine months later, she gave birth to the baby. It was born in a gush of water. "It was as if," he said, "the baby was made of water. The baby just melted away."

As he told us this, his eyes flitted nervously, as if the very mention of the incident frightened him. "When you are alone in the water and the boto comes, you are afraid," he said quietly. "They make a hole under the water—the Encante. And they really attract you. They seduce you. You must be careful—*cuidado com o boto*."

We next approached an old man with clouded eyes, working behind a towering pile of foot-long, shovel-headed catfish called surubim. I asked him if he ever encountered botos. Yes, he said. "When the boto comes, he takes away the fish from the net. Once I was so angry, I shot at one. When a man is very angry, he does a stupid thing." Fortunately, his shot missed. The boto has powers, he said, and then told us again how the dolphin comes to parties and seduces pretty women. A woman must be careful, he warned, especially if she has her period. The boto will enchant and seduce her, and she will remember it only as a dream.

Isabelle then told him that we had swum with botos, and that we planned to do so every day. *"Cuidado!"* he warned us, shaking a finger. "Why? Are we in danger of being enchanted?" I asked. Isabelle translated. "No, that's not what he's saying," she told me. "He says they are very big animals, and they bite people."

Of course, we knew the botos were capable of hurting us. Vera had warned us about their strength, and we remembered the boto who had rammed Roxanne, hitting her at the heart. Also, Steve Nordlinger had sent me a small news clipping earlier about Chuckles, the pink dolphin at the Pittsburgh Zoo. A zoo volunteer had forced her hand beneath the screening of his exhibit, and the dolphin, annoyed by this intrusion, had leapt up and bitten her finger. (She sued, but the court ruled the zoo was not liable.) When we later visited Chuckles, we learned he had also bitten several keepers, some of them rather seriously, in what the staff later considered may have been a territorial display during his rut.

We thought none of this surprising. But most people forget that dolphins, like all animals, have agendas we do not understand and under

some circumstances can be dangerous. At the marine mammals conference in Orlando where we had met Vera, Scottish researchers had reported on shocking attacks of bottlenose dolphins on harbor porpoises.
Unusual numbers of harbor porpoises began washing up on the
beaches, especially off the Moray Firth, between 1991 and 1993. Post-
mortems revealed forty-two of them had been bludgeoned to death.
The researchers discovered to their horror that, essentially, small gangs
of thug dolphins were beating them up. Ben Wilson of the University of
Aberdeen even had two videos of the attacks: Two or three adult dolphins would chase a single porpoise, who was clearly trying to escape;
the dolphins relentlessly butted their victim, sometimes sending the
poor porpoise flying clear out of the water. Wilson called the gruesome
movie *Jack the Flipper* and questioned whether "Swim with Dolphins"
programs were such a good idea.

None of this information, however, was going to keep me out of the
water. Simply being with the botos was worth a great price. If anything,
we were more eager than ever to see them that afternoon, because
both Isabelle and I were getting our periods. I wanted to test if the stories that menstrual blood attracts the dolphins were true. We might
also, I realized, discover whether there actually were any piranhas in
the water.

On the boat ride to Ponto de Curucu, as Dianne applied lip gloss and
jaborandi hair conditioner she had bought in Santarém, I swallowed six
aspirin to increase my blood flow.

But on that day, the dolphins did not approach us. We saw them in
the distance—perhaps five in all, fifteen yards or so away—but they did
not come near. I wondered if our previous luck was merely a fluke.

The following morning, when we went out to the point, Isabelle stayed
in town. She is fairer than I, and our previous outing had given her a
searing sunburn, which she was treating with one of her many herbal
salves. My skin stung, too: little blisters were forming on my arms and

chest, and my face was red as a uakari's. One night, I slept all night on top of a big wooden clothespin and never realized it, because my senses were flooded with pain from my sunburn. The itching that followed was even more annoying, especially once I located its source: hundreds of tiny brown ants discovered the bounty of my shedding skin and weeping blisters, and would flood into my bed each night to drink the fluid and collect the skin to carry it away. ("I hate it when insects eat my flesh," I had commented to Dianne. "Yeah," she agreed, "especially when you're still alive. It's OK when you're dead, though.") We decided that we should limit our visits to the dolphins to times when the sun was less strong, in the early morning and in the late afternoon.

On this morning, Dianne and I sat together on the beach, waiting for the dolphins to appear. I grew impatient and paced along the sand, scanning, expectant. No waves transformed themselves into faces and foreheads and fins. I walked farther, and finally walked right into the water, making myself an offering. There I waited, my mind drifting on the waves, willing a boto to appear. Then, suddenly, a shadow on the water—the play of clouds on the waves?—rose to the surface directly in front of me, and resolved into the dorsal of a boto. And then it disappeared.

Did I imagine it? At first I was dumbstruck, then desperate. It *must* be there, I thought. And then, again—close—there were two! Two dark fins rose at once, just ten feet away.

I remembered Dianne. We had promised to call out to the other if either saw botos. She was at the other side of the point, a thousand yards off, and my voice wouldn't carry. I started to run toward her, and as I rounded the point, I saw her waving her arms.

"Holy God, Sy!" She was breathless. "One came within five feet of me! And he *blew*!" It was so unexpected and so close, she said, that she was frightened. "I just stood there," she said. "I didn't know what to do."

We stood together waist-deep in the water. "We are in their element," I said. We sang for them, and splashed. Two again! One large, the other small—a baby born, perhaps, in June. They were ten feet away, then twenty, then thirty. Both leapt! The baby popped almost entirely out of the water. The face was so tiny, so perfect! Dianne's skin, I noticed, was covered with goose bumps.

A third dolphin appeared. All three leapt from a common center, exploding in different directions—a starburst of flesh. One lay on its back, waving a flipper. Another, a gray one, rose, not ten feet away, and raised his head out of the water and looked at me. I could see his eyes focus on mine. And immediately another rose beside him and, turning his flexible neck, stared into my face.

The dolphins seemed to be getting closer. But now I realized I was swimming, and the water was over my head. Dianne, still waist-deep in the water, called to me: *"You're too far out!"* I turned to see her, and she was only an inch tall, much farther away than I'd thought. I saw that the current had also carried me sideways and was about to sweep me around the point, where the current was much stronger. Reluctantly, I swam to shore.

"Jesus, I thought they were taking you away," Dianne said.

How willingly I would have gone with them. I had surrendered to their Encante. In the water, I was a creature transformed: no longer terrestrial, no longer bipedal, I shed the world of earth and air: I left behind the way I breathe, the way I move, the very weight of my body. Each immersion offered a baptism, a new birth. Inside the water, I swam in the womb of Mystery, where pink dolphins looked into my face, where every element, including my own future, was utterly out of my control, and where the most impossible of possibilities come true.

As I emerged from the water, my body suddenly felt leaden, spent.

DANCE OF THE DOLPHIN

"Our legend is that there was a very pretty girl in one family. Here, you see, we used to hold a ball—the Rose Ball—and before the dance, the girl came to bathe in the river. She didn't notice she was being watched by a boto. But from that time on, the boto wanted her."

Through Isabelle, Necca is telling us the story she learned from her grandparents. Necca's full name is Ludinelda Marino Gonçalvez, and she is forty and wiry, with the high cheekbones of a Bourari Indian, a heritage of which she is proud.

Even though we hear her words in translation, as we sit beneath the thatched roof of the cabana where she sells fried fish and soft drinks at the edge of Lago Verde, we can see Necca is a master storyteller. She reveals her story slowly, slyly. As she speaks, her eyes slide from corner to corner, as if spotting some detail there to remind her of the next turn of events.

"The night of the ball," Necca says, "the beautiful girl arrived with her boyfriend. But the boto was the handsomest man there. She looked at him and fell in love—and he with her. They were blind for one another."

Oh yes, Dianne and I agreed with a glance. We knew that feeling: when love floods the senses, jams your sonar, blinds you to all else. Lightning might crash around you, eels and piranhas nibble at your

toes, and you don't care, because only One Thing matters—that longing which has overtaken your soul. We humans were made for that
sweet, sweeping sickness. But what if you have fallen blindly, impossibly in love with a dolphin?

Necca continued: "One man realized that this was a boto. He chased
him away. But the girl loved him. She loved him madly. At the next
ball, she came and looked for him. There he was, and they danced all
night in each other's arms, and then they walked on the beach.

"They lay down in the sand, and there they made love. They did
everything with each other," Necca said, inviting us with her eyebrows
to imagine the details. But after the lovers fell back into the sand exhausted, he suddenly leapt up and disappeared into the water.

"She was heartsick. She went to the village *curandeiro*, hoping he
could help her find her lover. The *curandeiro* asked help from the
Mother of the Lake. The Mother agreed to ask the Moon to call him
back. For the girl was now pregnant. And the *curandeiro* saw that the
father of the child was a boto.

"At the next full Moon was the next ball. The Moon called to the boto, and he came to the ball to meet his love. And then the girl told him she was carrying his child. Now he was forced to explain: though he loved her and longed to be with her, he could see her only at the balls, for he could change into a man only on those nights.

"It was months before the next ball. When they met again, she brought him his son. Every ball thereafter he came to see her, to dance with her and see their son."

Necca choreographed a dance to tell the story. Braulio had told us this earlier: the dance is performed each year, the last week in July, at a festival at the water's edge in Alter do Chão. Necca has danced in it, and now her beautiful twenty-four-year-old daughter, Keila, performs; when Keila's young daughter is old enough, she will dance, too.

"Many people, old people, believe this story," Necca told us. "Many girls today come to the balls hoping to meet a boto! For they are the handsomest men there. And when people come to see the dance, dolphins often come near the beach, as if they knew the dance was about them."

As rain was our companion on earlier journeys, now our companion is fire. We had welcomed the rain, flooding the world with its brimming abundance; but though we dread the fires, we cannot escape them. Some days, on the horizon at Curucu, there are four or five columns of smoke; some days the air is leaden with it, pressing the river flat as quicksilver. One night, fire had nearly come into Alter do Chão itself, so close we could feel its gnawing heat. Always, somewhere, there is fire or smoke, insistent reminders of the greed consuming the world.

As we head to Curucu this morning, we see three big fires to the west, and the air is hazy. Only in the river can I find respite from the burning. As Dianne unpacks her photo gear, I swim out to meet the first fins. I

Storyteller Ludinelda Marino Gonçalvez told us ancient Bourari legends.

breaststroke slowly and they come to me: first, the mother and baby; one very large pink adult; two big grays.

I am surrounded, but I do not know this; Dianne, taking photos from the top of the *Gigante,* later tells me there were dolphins all around me. Two pinkish tucuxis joined the group, one with a slash across the dorsal. For three-quarters of an hour, there was not one minute that I could not see a dolphin at eye level. As they surfaced, it seemed I could feel their glance. Or perhaps it was their sonar. Later, I would learn that people who work with marine dolphins say they feel the animals "sounding" them, throwing out trains of ultrasonic clicks created by

moving muscles in the melon, and waiting for their echo to return. Their sonar retrieves a three-dimensional soundscape, a sonogram, of what lies ahead of them; the boto, with its flexible neck, can turn the head in a wide arc and obtain an exceptionally broad sounding. One researcher, Bill Langbauer at the Pittsburgh Zoo, told me when the sound waves hit him, it feels like humming with your teeth clenched.

But that was inside an aquarium tank. I swim inside of living water, and always I feel the water humming, charged with life like the blood in my veins. I cannot feel the botos sound me, but surely they are doing so. They can see through me, as God does. They do not touch me, but their soundings penetrate my flesh. They know my stomach is full and my womb is empty; they can see the faulty mitral valve in my heart. And yet, even possessed of this astonishing sonar, they still pull their sleek faces out of the water to look at my face. Why would my face be important to them? They can recognize me by the other nine-tenths of my body beneath the water. Yet they look into my face again and again. They must know we humans wear our souls on our faces. Perhaps, to them, too, a meeting is more profound when it is face-to-face.

And what do I know of them? I only know the sex of one—the mother with her baby. I do not know how long they have lived. I do not know how far they travel. I do not know if this group always stays together, or only comes together here to investigate the pale, terrestrial stranger who hangs in the column of water before them.

But somehow I am not frustrated that I will never learn the answers; this is not, I now know, what they have to teach me. Already, their kind had shown me so much: the botos brought me to Manaus, the impossible Paris in the Amazon; the botos drew me to the Meeting of the Waters. The botos drew me to Tamshiyacu-Tahuayo, and to Mamirauá, and now, to the clear waters here. Throughout these four journeys in the Amazon, I had followed them, though not in the way I had origi-

Burnt landscape in Pará.

nally planned. For the verb "to follow" carries many meanings, most of which I hadn't been aware of when I had decided to follow them. To proceed behind them, to go after them in pursuit, as I had envisioned when I'd hoped to trace their migration, was merely one way of following. Other meanings are more subtle and profound: to be guided by; to comply with; to watch or observe closely; to accept the guidance, command, or leadership of; to come after in time; to grasp the meaning or logic of. And so, without chasing them along a migration, without a single hit on Vera's telemetry, I had followed them nonetheless. In Tamshiyacu-Tahuayo, I had followed them to the spirit realm, where shamans commune with the powers of the plants and visit the Encante; with Gary, I had followed them back through time. At Mamirauá, they had taken me to the heart of the Amazon's modern conservation dilemma. And now, I followed them still, hanging in the water before them, ready to receive their gifts and their guidance.

I always save the moment I enter the water for them. Dianne and I wait at the edge of the river, and I do not swim until the dolphins come. I spot one fin, close, and give my body to the water.

Dianne stays at the edge of the river, or sometimes on the roof of the *Gigante*, photographing. She is a strong swimmer, but her father was a sea captain; those who know water most intimately, I have found, tend to more wisely respect its dangers. But I lose myself utterly in the water, like a soul leaving a body.

Now that we have come here daily for a week, we recognize seven individuals: there are two medium-sized grays, the bowler-hatted businessmen. In fact, Dianne knows an Englishman who owns an actual bowler hat, and we named one of the dolphins after him—David. The other we named Gary, whose distinctive rain hat had been the envy of everyone on my second trip to Peru. One very large gray dolphin we

The author swims to join a dolphin in the Tapajós River.

name after Vera. A very large pink one we call Valentino. Another is gray with pink lips, and we name him that—Pink Lips. And occasionally we spot a mother and baby. It is nearing Christmastime, so we call them Mary and Jesus.

Pink fins, gray bellies, the bulbous soft heads; I so wish I could touch them. I can feel the currents their bodies make as they slip through the water. One day, I look down into the water beneath me and see a large gray form swim under my feet. All but immersed in these clear blue waters, I feel as if embraced by paradise.

In Curucu, we had found an ideal site. We could imagine no situation better. But we still felt we should visit the dolphins of the Arapiuns.

At the Mercado 2000 in Santarém, we had met a fisherman with thinning hair and strong hands named Valdomiro Ribero who had spent a month on the Arapiuns when he was fifteen. He would never forget it. "There were many, many dolphins there," he had told us. One night when he had slept on the beach, he had spotted two strange-

looking girls, perhaps eleven and thirteen years old. Their skin was very white, he said, but had reddish spots. And they had very little hair. These girls, he had been told, were the boto's daughters.

Certainly, there had to have been a reason that film crews had by-passed our lovely Alter do Chão and made the day-long journey up-river to the Arapiuns to the west. Perhaps, Dianne and I thought, the waters were even clearer there, the dolphins more easily seen.

We found making arrangements for the journey difficult. Not every boatman will risk it. Braulio told us his *Gigante* could not negotiate the current. The portion of the Tapajós leading to the Arapiuns is difficult to navigate—there are dry spots and rapids—and besides, he said, the place is full of mosquitoes.

It took us five days to arrange for a boat. Sadly, Isabelle was leaving us, to visit Brazilian friends in Belo Horizonte, but she helped us get the trip organized before she had to depart. Socorro knew a boatman, Gilberto Pimentel, who agreed to consider the trip. Necca's beautiful daughter, Keila, speaks some English, and she agreed to help translate our negotiations for the boat. Grateful for Keila's and Socorro's help, we invited them to accompany us. To our surprise and delight, they accepted.

Through Keila, Gilberto explained that his boat was a strong one. It is named *Boanares*, after his father, who accompanies him on his trips. "But even with a strong boat, the trip can be dangerous," he warned. "Many boats break up when the water is angry, and there are many shifting sandbars. The river is so wide that you cannot swim to shore." We should plan on bringing supplies in case we got stranded, he told us. In any event, we would need to spend one night on the water, where, he confirmed, there would be many mosquitoes.

We left at four in the morning, when the wind was very still. But af-

Keila Marinho carries on Bourari tradition.

(Photograph by Sy Montgomery)

ter an hour of travel, the waves grew so rough that the little table in the cabin fell over, and all of us but me were thrown from the hammocks we had strung up in the boat's central cabin. We sat on the floor. We stopped for an hour at Ponto des Pedres—Place of the Stones—to wait for the water to calm down.

Another hour's travel upriver, at about fifteen miles per hour, and we reached a spot called Icuxi. Here, Gilberto strung up his seventy-foot-long net, with plastic Coke bottles for floats and a brick of red sandstone for weight. Once the net had caught some fish, he explained through Keila, the dolphins would come. So we waited. We lay in our

hammocks. We walked along the beach. We waded in the water. But, although the net caught several fish, no dolphins came.

We continued on. We passed more white sand beaches. We passed innumerable fires. Finally, toward sunset, we reached the Rio Arapiuns. Its blue waters were calm. No fewer than nine fires burned on the western horizon. To our surprise, the Arapiuns looked exactly like the Tapajós.

As night rose, we strung up our hammocks again, and Dianne and I fitted ours with mosquito nets. The mosquitoes were as voracious as predicted. To our dismay, we discovered that Socorro and Keila had not brought netting, but our friends did not complain; clouds of Portuguese were rising from their hammocks in an animated discussion. I imagined their Portuguese somehow repelled the mosquitoes, like a column of smoke from a mosquito coil.

I listened, and to my surprise, even with my rudimentary grasp of the language, I found I could understand their conversation:

"Do you think a boto will come?"

"Will he be enough for four women?"

"And we'll need one for Gilberto! And for his father!"

"Oh, we will not sleep tonight! We will wait for the boto to come!"

"If only we had festa music to attract him!"

Rocking in my hammock, I could feel the waves rising beneath me like a lover.

We woke at five and headed to a place called Jari. There were many shallow channels there, and good fishing—for both people and botos, Gilberto said. With Keila, Gilberto took a rowboat out to speak with the fishermen, to see if they had seen any dolphins.

Yes, they said—right over here! They pointed to a channel separating two grassy shores where skinny cattle were grazing.

And sure enough, we spotted three botos: a large pink adult, a large gray adult—and a small, grayish pink baby.

Quickly, Gilberto and his father set the net across the channel, confining the botos to the shallow waters where we could best see them. The water is only waist deep and somewhat muddy. Dianne quickly loads her camera. But in the channel, something has gone wrong. Something is caught in the net—the baby boto! The bright pink adult starts to thrash in the water, showing her flippers, obviously distressed. The baby might drown! I jump into the water—without thinking to remove my sneakers—and immediately sink into the mud up to my ankles. Gilberto and his father jump in, too, wisely barefoot, and rush to the net. I stagger to join them.

The baby is squealing in terror, and with every effort to escape, entangles itself further. Gilberto's father, Boanares, reaches it first. I try to hold the baby still and keep the head above water while Boanares disentangles the flippers and tail. He does not want to tear the net. A seventy-foot net takes more than a week to make by hand, and in a store costs $150. "Cut the net! Cut the net! I'll pay for a new net!" I bellow, but of course Boanares doesn't understand me because I have forgotten all my Portuguese.

The baby screams in my arms. Its melon heaves with the force of its terror, its blowhole opening and closing, howling like a tiny mouth. The infant is nearly four feet long, and shockingly strong, one big muscle tensed in panic. The best I can do is hold it still, try to calm it down. I slowly stroke its skin, running one hand under the belly, holding its side against my pounding heart. The skin feels like a boiled egg, and is now flushing pink with exertion. I know now the baby will not drown, but I fear the net will cut the delicate skin. I look into its pearly eye. To the terrified infant, I am a lightning storm, a caiman, a pack of piranhas, an evil spaceship. The baby has no way to understand that we are trying to help.

Finally, Boanares frees it from the net. I hold the baby for a moment longer, at once a heartbeat and forever, my clasp around its body a plea and a prayer. And when I release the boto, I let go my heart from my throat, and feel the water surround my empty arms like forgiveness.

But now the mother is caught in the net! Dianne hands her camera to Keila and gets in the water with us to help. The mother is much calmer than her baby. Dianne wants to hold her, and the mother doesn't struggle in her arms. I almost wonder whether, in the manner of mother birds feigning a broken wing, the mother had embedded herself in the net to distract us.

Now the dolphins swim free. The dark one leaps, as if in triumph. Gilberto and his father remove the net. We all adjourn to the orange and blue *Boanares*.

Everyone is exhausted. Socorro and Keila had been terrified that the two adult dolphins would attack us while we tried to disentangle the baby. Gilberto and his father were afraid that somehow the Brazilian environmental police, IBAMA, would find us with a dolphin in our net and arrest us all. And Dianne, though seeming cool and poised, was worried, too: she was afraid that she would miss the shot. But she didn't.

I collapse with emotion—the thrill of holding the baby, the terror that it might be injured, the guilt that had a boto been hurt, it would have been my fault. In the Tapajós, I had so wanted to touch the botos. But here, we had not intended to catch the dolphins, only to briefly confine them; yet this near tragedy, which could have drowned two botos, was, I was sure, a consequence of my own desire, flaming like a fire on the edge of the horizon. Again, I am reminded of the damage an outsider can wreak.

My sneakers are encased in mud. My feet are made of clay.

On the way back, we stopped at Curucu. Even after only one day away, we missed "our" dolphins, and although Keila and Socorro knew we were swimming with them daily, they had never seen such a thing and were eager to watch.

Within three minutes of our arrival, the dolphins appeared. It felt like

Fires on the horizon.

coming home. I swam out to them, perhaps a quarter mile. All seven botos appeared, blowing, pulling their heads from the water to look. A single tucuxi leapt. Valentino surged out of the water, showing his flippers. One of the medium grays, Gary or David, rolled on his back, waving his flippers, and then turned, flipping his tail.

Meanwhile, Gilberto and Boanares set the net beside the boat. Within an hour, it was heavy with fish. Dianne, Socorro, and Keila stood on the roof of the *Boanares,* watching—and suddenly, fast as an arrow—*"Rápido! Rápido!"* shouts Socorro—Valentino shot by the net upside down and plucked a fish from the mesh. Another dolphin— which I couldn't see—tried to grab it from his lips. Valentino rocketed to the surface, shaking his head with the fish in his jaws, and jetted away.

We waited to see if more dolphins would feed from the net, but none came. When we retrieved the net, we noted with horror that it contained two four-inch, red-bellied piranhas.

· · ·

Periodically, we would take the bus into Santarém to change money. It was usually an enjoyable ride of an hour and a half or so, depending on the route, and we became such familiar passengers that the robust, mustached bus driver began to greet us (particularly Dianne) with a hug. Brazilians seem to try to fill every silence with music, and like most buses, this one came equipped with a tape deck. Even though we would have preferred Caribe or boi-bumbá, our driver often played American artists in our honor. On this December day, as we return to Alter do Chão from some morning errands, he plays an Elton John Christmas tape. Outside, it is snowing: huge black snowflakes swirl through the windows, the cinders of a fire whose heat we can feel nearby.

We are driving through thick smoke. Our driver is unperturbed. On one side of the bus, the trees are burning. None of the passengers seem concerned. An entire side of the road is flaming like a Yule log, and absolutely no one cares—until we come to a corner where one of the fires chews through the thatched roof of a house. Now they gasp and cry, "*A casa! A casa!*" and crane their necks to see.

When we return to Alter do Chão, the skies are smoky and brooding. The day before, we had learned, there had been such a dangerous wind on the Tapajós that many fishermen raced home early. But today the wind is even stronger, as if fed by the force of the flames.

We have a larger boat today. Braulio, too, had to go to Santarém on an errand, and arranged for his friend Manuel to take us out on his forty-foot boat, the *Sabiá*. The craft bucks on the waves. We anchor at Curucu, and immediately Dianne spots a pink fin. I go to him, swimming out a quarter mile before I stop to look, treading water.

The waves slam against my face. They jostle, like tall people who crowd in front of you at a parade. I try to peer around them, only to be slapped again and again. The water goes up my nose and down my throat. I am swallowing the Tapajós, and it is swallowing me.

I hear a blow. I turn to spot two grays. David and Gary have surfaced

View from up the machimango tree. Below, people are fishing for piranhas.

While paddling his canoe, Don Jorge Soplin was nearly seduced by a dolphin. (Photograph by Greg Neise)

Several fishing spiders, Moises told us, worked all night to create this giant web, which stretches as long as a badminton net.

Below, left: A hairy megalomorph spider, one of Dianne's few fears.

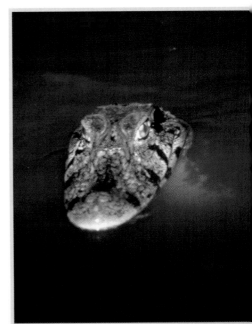

Giant lilies as big as throw rugs at Tibe Lake.

Below: A dead caiman.

Below, left: Sometimes the lakes were so crowded with caimans that our flashlights lit a wall of glowing eyes.

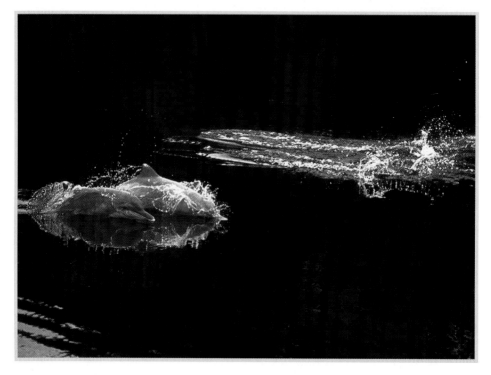

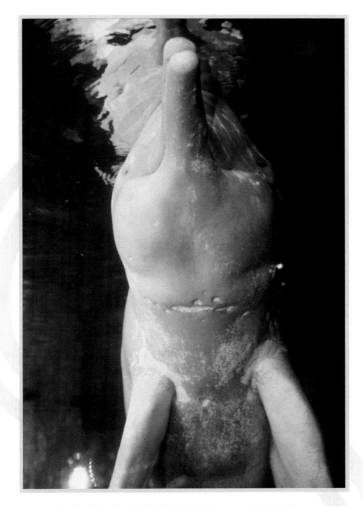

Large and winglike, the pink dolphin's flippers contain the bones of five fingers, like our hands.

Above, left: A pink dolphin at Pacaya-Samiria in Peru. (Photograph by Thomas Henningsen)

Bottom, left: The prominent dorsal fins of the small tucuxis make these river dolphins easy to distinguish from pink dolphins, though both species can look pink. (Photograph by Thomas Henningsen)

Top: Pink dolphins tend to hunt singly, and may dispute fish in splashy fights. (Photograph by Thomas Henningsen)

Bottom: An exceptionally brightly colored pink dolphin displays the low dorsal ridge of this species. (Photograph by Thomas Henningsen)

Top: Baby pink dolphins seem to like to swim touching their mothers for comfort. (Photograph by Thomas Henningsen)

Bottom: Pink dolphins' color can vary enormously, from pearl gray to charcoal, from deep rose to pale pink. (Photograph by Thomas Henningsen)

Sunset transforms dancers to dolphins on the beach at Curucu.

next to each other, seven yards from me. The sound of their breath comes to me on the wind and seems as if it is in my ear. I feel safe with them.

But meanwhile, Dianne grows nervous. The two-foot whitecaps whack the boat so hard that bottles and fishing knives crash through the cabin, a bar of soap skids across the floor, a stool flips over. From the poop deck of the *Sabiá,* she can no longer see my head at all, not even with her camera's 300 mm lens or her binoculars. She calls to me: "*You're—too—far—out!*" but the wind takes her voice away.

I'm treading water, scanning for dolphins. The waves behave like a schoolroom of naughty boys: they grab my hat, snatch at my glasses, pull at my bathing suit. A forehead surfaces and dives, a dorsal rises and rolls, the waves bob up and down—it's like a hall of mirrors, the dolphins and the waves, the waves and the dolphins, until I feel as if I am staggering, drunk. And then the dolphins are gone.

My arms and legs are working hard to fight the current. I feel the water trying to sweep me around the point. I look back to see where I am and notice the boat looks small as a toy. And now Dianne's voice comes to me on the wind. I work my way back to shore—a tiring swim. Dianne jumps out of the boat to meet me in the water. I've been inside the waves for thirty-five minutes, and as I emerge, I realize my arms and legs are tired. But just as we sit down on the shore together, we spot another pink fin. I go back out.

A tucuxi leaps in front of me. Dianne tells me later that half a second after that, two other tucuxis ganged me from the back. Most of the time I couldn't see them, but I was surrounded by dolphins—tucuxis and botos, like guardian angels against the storm. Perhaps that is why I was never for a moment afraid.

I come back to shore and together, Dianne and I watch the smoke-stained sunset. Just as we are about to leave, we spot big, pink Valentino. Amid the waves, his dorsal ridge seems to hang above the water like a sail in the air—weightless, timeless, impossible.

. . .

One afternoon, Braulio failed to meet us at the *Gigante* at the usual time. While Dianne waited with the photo gear, I trekked back over the sand to his house and found a spiffy red car in the courtyard. Braulio was next door at the restaurant, the Lanchonete, with a young, dark-eyed lawyer named Felicio Pontes, Jr. I invited him to join us. He speaks French, Portuguese, and English—the English with a French accent. He lived in Paris for a year. Born in Rio, he worked in Belém for a year before coming here to argue cases to preserve the forest in Pará.

Unlike the local Brazilians, Felicio swims and swims well. But he won't go out into the waves even a third as far as I do. He and Dianne wait as I swim to the dolphins—five of them today. Valentino surfaces close to me, opening his blowhole not five feet away, intimate as a kiss, and then slowly rolls over. David and Gary surface side by side, then dive. I can see Mary and Jesus in the distance, stealth-floating near the surface. They are eyeing the stranger among us cautiously, I imagine. When Mary dives, she shows the flukes of her tail.

The others stayed near for half an hour and then swam away. I returned to Dianne and Felicio, and as we stood together in the clear water, Felicio told us Indian legends. He told us how the tears of the Moon, weeping for her lover the Sun, had formed the Amazon, the river of impossible love. He told us how the tambatajá tree was born: when a Macuxi warrior's wife was killed in war, the grief-stricken husband buried himself alive beside her grave. The beautiful bananalike tree arose from that soil, he said. "It is a plant very *melancólica*," he said, "very huge with branches like arms closing."

How was it he came to know these stories? I asked. "It's important, in order to convince the judge of your argument, to know what is important to the people," he answered. "I prefer always to stay with the people and learn how they understand the world."

As an environmental lawyer, Felicio represents the people who still swim in the tears of the Moon, who still see their own love and tragedy twined in the arms of the trees. Felicio tells us of one case he recently argued and won: "There are a hundred and ten falls in the beginning of the Tapajós River," he says, "and they were going to explode them, to make the river more navigable for commerce. But the Indians there told us this was happening—and we stopped it.

"It is very recent, this kind of justice," he says. "Belém was the site of a federal court, and now, in 1996, it begins in Santarém." For today the courts are the stage for the operatic clash of different worldviews: on one side, extraction and commerce; on the other, wholeness and holiness. And it is a drama, we learned, played out in the river in which we swim—a drama that continues now, even in our flesh.

That night, we invited Felicio and Braulio to join us for dinner. As we shared the delicious tambaqui com leite de coco that Socorro had prepared, Felicio told us that the Tapajós was a river remade.

"Two years ago, you couldn't have seen your dolphins here," he said.

"Why not?" we asked.

"Because," said Braulio through Felicio, "the Tapajós looked just like the Amazon—all brown and muddy!"

"What!?" Dianne and I were astonished.

"There was no difference!" chimed in Socorro. "Everything was contaminated."

The river had been the victim of an Amazonia-wide gold boom that during the 1980s had rivaled the California gold rush of the nineteenth century.

And what a boom it was: At its height here along the Tapajós, in a single day more planes would take off and land in Pará's tiny, regional Tatuba airport than anywhere else in Brazil. People walked around in the streets carrying gold in their T-shirts, stretching out the fabric like baskets, Socorro remembered. "And at that time, you bought things with gold and not money," said Socorro's husband. He had bought a certain motor for less than three ounces of gold back then; today that motor costs $3,000. "The price of the machine has gone up, but the price of gold has gone down," he explained.

In the last decade, Michael Goulding has written, gold has been "the single most valuable resource exported from the Amazon basin," earn-

ing revenues estimated at $1 to $3 billion. The reason for this wide estimate is that as little as a third or even only a fourth of the gold extracted from the Amazon is sold through official channels. Most of the
miners—an estimated half a million of them in the eighties—are inde-
pendent fortune-seekers, called *garimpeiros,* working in the rivers illegally for so-called "placer" rather than "hard rock" gold.

Back at the Meeting of the Waters, when we had begun our journey,
Dianne and I had met one of the men who risked his life for gold. He
spoke English and introduced himself as Raymond Des Costa, age
thirty. He was wearing a Reebok muscle shirt and a black cap advertising a Pink Floyd album with the phrase "Up Against a Wall." Raymond
and some friends had just come from Colombia to Brazil looking for
gold-diving jobs. It is not an easy life, he explained: "For two months I
trained, and then for two weeks I suffered." On one of his first professional dives, he had gotten the bends, the painful bubbling up of nitrogen in the blood, and was hospitalized for two weeks, bleeding from
the nose and ears.

The divers first check for gold at the river bottom, eighty to ninety
feet down. If you find any, he explains, they send in "the Missile"—a
suction pump with hoses operated from a barge or raft. It sucks up the
river bottom, then spits everything out. "The Missile, it can throw
down big trees, eat all the mud," Raymond said, his eyes flashing with
fear and excitement. "If your hands go in its mouth, it will break them
up. If your belly go near, it sucks it out. If you be around, you be dead
when the motor starts hauling. Thank God I live." The money is good,
he said. But he lives a dangerous life, as one can see from the tattoo a
girlfriend inscribed on his arm eleven years ago: a cloud with his initials
circles a heart with a dagger through it.

Thousands of these dredges sucked greedily at the sediments of the
Tapajós, clouding its waters with mud. Now we understood why the
film crews had all gone to the Arapiuns; for a decade, the waters here
were too muddy to film.

But the worst damage was unseen. To extract the gold from the sediment, *garimpeiros* add metallic mercury to the deposits in their sluices. Much of the mercury falls directly into the river. Some of it adheres to the gold, isolating the precious ore. The two metals are then separated by blowtorch; the mercury evaporates and leaves the gold behind. But the evaporated mercury does not disappear. It falls as toxic rain into the forests and rivers.

A report of a United Nations research project called mercury pollution from gold mining "one of the most serious environmental problems in Amazonia today," warning that "damage from gold mining in Amazonia may be felt for decades to come." Predatory fish such as tucunaré, pirarucu, aruana, piranha, and many catfish—the species most people here eat—are most likely to accumulate mercury rapidly. "The long residence time of mercury in river sediments can contribute to health hazards long after the gold mining frontier has moved on," wrote its four respected authors, Nigel Smith, a professor of geography at the University of Florida who has worked in the Amazon since 1970, and his three Brazilian colleagues. The consequences? As much as 10 percent of the mercury one ingests lodges in the brain. Insidiously as greed, mercury clouds human perception and response to the world: as nerve damage progresses, it eats away at movement, hearing, feeling, speech, and thought.

I would read the U.N. report after my return to the States. To my horror, I would learn that in four communities tested in the Tapajós watershed, over 60 percent of the people had mercury levels in their urine high enough to warrant regular testing as recommended by the World Health Organization. Over half the river sediment samples taken from Tapajós and its affluents exceeded the limit for safety set by the Brazilian regional secretariat.

Gold-mining camps, the report further noted, transmit diseases. Miners introduce new strains of malaria to local populations that have no resistance. Indians are likely to suffer more severe malaria symptoms if gold miners are operating nearby. In Roraima, the new frontier to which more than 50,000 hopeful miners migrated by 1990, hundreds of Yanomami Indians died after the miners arrived. In one two-week period in 1992, 44 Yanomami died from malaria in one village alone.

But as Michael Goulding notes, the early-stage symptoms of mercury poisoning are very similar to those of malaria: fever, chills, nausea. It is possible that the Yanomami literally died of gold fever.

Felicio takes consolation in the Tapajós's latest transformation. "In two years, we can reverse the situation," he said. By 1992, most gold miners began to leave the Tapajós for new frontiers. Today, with the price of gold less than half what it once was, the water here is now clear as the diamonds in Dianne's dream. The river looks as if it has completely recovered. No wonder we consider water an almost infinitely forgiving medium: with it, we try to wash our sins clean.

I felt heartened by Felicio's optimism. I was grateful that the Indians and their environment had this kind, wise, energetic young man to speak eloquently on their behalf.

But in the sediments of the river, in the sweet flesh of the tambaqui

we ate that night for dinner, in the bodies of the dolphins with whom we swam, in the organs of the people we were coming increasingly to love, and now, in tiny amounts, in Dianne's brain and mine, the mercury lingers still.

Necca was well aware of this when she started her dance troupe in 1993. People were forgetting the old ways, she said. Many even forgot that Alter do Chão was once originally called Aldeia dos Bourari— Village of the Bourari. And when people forget who they are, they forget how to act. Look at the fires, she said: Sometimes they burn for four days! Night and day they burn, and she cannot sleep for the anger. And look at the garbage on the beach: everywhere, as we had seen to our dismay, the white and lavender sand was littered with plastic bags, drinking straws, jugs. "They spit in the plate they eat from," she said. "The people don't understand. They forget they depend on the land."

But once, her people knew. Once, she said, they could talk with the Moon; the Moon was the goddess of growth, protecting pregnancies, manioc gardens, forests. You must never cut wood at night, she said, for the Moon is jealous of her trees. This the Moon remembers, even if the people forget. And then Necca told us this story:

"One day," she began, "the fishermen of the village noticed that the most beautiful girl in the village had disappeared. They couldn't find her anywhere! So they asked the *curandeiro* to get her back. He performed the rituals to ask the Moon for help. Where had she gone?

"The Moon answered: she had been enchanted into the lake. But the lake, the Moon promised, would give her back. The people would see her again.

"At the next full moon, there was thunder. And in the middle of the lake, there appeared a beautiful tree, glowing brilliantly like the Moon,

"Walking Tree" at Lago Verde.

and with colored fruits. The tree went for a walk out of the lake. The people were surprised and frightened! Then the tree came back and sank back to the bottom of the lake. Its fruits fell—and they changed into the beautiful green frogs who gave their color to the lake." The lake was once called Lago do Muiraquitã, Necca tells us, in honor of the little green frogs there, who bring luck.

"The beauty of the Indian girl became the beauty of the lake," she explained. "And on full-moon nights, the tree would rise from the lake and walk about the village.

"But one day, she didn't come back from her walk," Necca said. And then she was silent.

"What happened?" I asked.

"We are still waiting," said Necca, "for her to come back."

And as they wait, the world is falling apart. "There is a danger," said Necca, "because our sons and daughters may never see the world as nature intended." They may never learn how to build a grass house. They may forget the herbal medicines. They may forget how to make pottery, charcoal, natural dyes.

There is trash on the beaches, fire in the sky, mercury in the water. "There was the epoch of rubber," said Necca's younger sister, Laurenice, who had joined us at the little table beneath the thatched cabana on the beach, "and then the epoch of gold. Now it is the epoch of wood— and even that leaves speaking English. And what does the villager get? Diseases. Mercury. There will be no dividend for the people."

The dividend for the people will not come from rubber or gold or timber, Necca said. It will come with remembering. Necca and Keila dance to remember the wholeness of a fractured world.

"I wish that you had come in the summer," Necca said to us. "You could see the Dance of the Boto." We wished very much we could see

This poster advertises a reenactment of the legend
of the enchanted dolphin.

it, I said. We wished we could stay longer. But Christmas was coming, and friends and family expected us back in the States. In a few days, all too soon, we would be leaving.

The morning of our last day in Alter do Chão, we visited the Indian museum.

WE ARE NOT A MUSEUM, a placard informed us in English as we entered the door. WE ARE THE CENTER FOR THE PRESERVATION OF INDIGENOUS ARTS, CULTURE, AND SCIENCES. Its mission: "To document and preserve for the future all that has been forgotten, discarded, and overlooked in

the indigenous history of the Amazon." An ambitious agenda, Dianne and I thought.

The Center, we read, opened in 1992, "the great dream of Maria Antonia Kaxinawa," David Richardson's wife. We found her stringing beads for the museum's gift shop. Similar-looking items—pottery with frog and geometric designs, basketry, beadwork—are available in Santarém, but a sign in the museum decries such merchandise as "airport art."

As you enter the impressive yellow cement building, you are greeted by a large central mound of vegetation, which reminds me of a waterfall, and in fact there is a pool at the bottom. This display is dominated by a large inscription:

After I had taken a lot of the drink, I started to chant. The ground became red and flattened, beautiful. The sky began to sing We! We! We! The colors of the rainbow began to appear and swirl about like a snake. The spirit allies began to arrive. . . .

My soul began to shine.

One by one, the spirits arrived—the MOKA—the frog spirits with quivers of arrows on their backs. The peccary spirits . . . the spirits of the waterfall and the fish spirits. All game had moved into my chest. . . .

I was deeply moved by the inscription. I asked Maria Antonia where the quote came from. "Oh, it's something David got out of a book," she said. She knew nothing more about it. Later, I read the inscription to Mark Plotkin, who has studied Amazonian tribes for more than fifteen years. He said it almost certainly described a Yanomami ceremony in which the people snort a hallucinogenic snuff, the way others take Ayahuasca, to contact their spirit guides.

I was disappointed to discover there was nothing to be learned in the museum about the Bourari. "They are completely acculturated now,"

Maria Antonia said, so the museum pays them no tribute. Instead, the center concentrates on displays of art and artifacts of tribes now extinct or nearly so. Among them were the Tapajós. Before the arrival of the Portuguese in 1626, this river basin, we read, was the economic hub for the Amazon. The Indians who gave the river its name had built an impressive civilization here. Eight thousand years ago they created great art, especially ceramics, including large, ornate funeral urns. The craftsmanship of their pottery "surpass[es] the finest Venetian glass," wrote an explorer. But the Tapajós people went extinct, succumbing to foreign diseases, a mere forty-seven years after first contact.

More than ninety Indian tribes in Brazil alone have become extinct since the turn of the century, a rate of one tribe a year. Murdered and enslaved, victims of foreign diseases, greed, religions, and alcohol, they are disappearing still. Mark has spoken eloquently of this loss: when one of the medicine men of these vanishing tribes dies, he has said, "it is like a library has burned down. Only it's worse, because the knowledge in a library is recorded elsewhere, and when these men die, their knowledge dies with them."

The Center is like a library in ashes. For each tribe, there is a display of artifacts: woven basketry, feather headdresses, masks, and sometimes photographs, like items rescued randomly from a burning building. Below the objects, handwritten in Magic Marker on cardboard cards, brief texts in four languages give a fact or two about each lost or dying culture. The words on the cardboard placards are succinct as tombstones:

The Kanamarís: The first explorers to contact them in 1940 found a culture rich in song, dance, and the art of permanent facial tattoos. Now there are 643 of them left in the state of Amazonas. "In 60 years," we read, "they traded their past for alcohol."

The Yanomami: Known as the "Fierce People," 62 percent of them tested positive for new strains of malaria brought by gold miners, we read, and only about 8,000 of these people now survive in Brazil.

The Waimiri-Atroaris: Their territory, in what is now Amazonas state, was once the most feared in the Amazon. Their population decimated to under 3,000, they surrendered to pacification in 1977 to make way for the Pan-Amazonian Highway and a hydroelectric dam.

The Matses: Their facial ornaments include bones through the nose to resemble whiskers and shell earrings to pay mystical homage to the jaguar. Masters of the blowgun, their hunters could kill a hummingbird in flight and a monkey at 130 feet. Since first contact with whites in 1976, their population has dwindled to 123.

The Tukanos: Their legends once told how the first man found a sacred trumpet, and from its music flowed the stars and the wind, the rivers and fish, the forest and game, and all his children, including the tribes of the Desana, the Wayanas, and others. The 2,631 who remain are now dominated by Christian religious groups.

The Asurini: Native to the Rio Xingú, they were known for the fabulous geometric designs they painted on their pottery and bodies. There are only 10 women and 7 children left.

The Tenharim: Players of the sacred flute, they adorn themselves with beautiful feather necklaces and earrings, and crush dried flowers to sprinkle them on the body after the bath. Fewer than 350 are left, many struggling with alcoholism.

The Marumbas: Native to the Vale do Javari, they were decimated by the slaving and murder of the rubber trade. Only 622 survive.

Culture after culture is rendered down to a sentence or two: the Campa, master weavers who brought the pan pipe to Peru; the Desana, who called themselves "Sons of the Wind"; the Waiwai, who play sacred jaguar-skin drums; the Wayaná/Apalai, who believe a mystical bird created the world. But today their world is on fire, consuming their songs, their knowledge, their language, their forest, their gods.

Of the people called the Parakanás, we learn only two facts. They once lived along the Xingú River basin and in Pará. There are only 300 of them left. The rest of the text on their card asks a question: "We can wonder what the world would be like without the magic of the mighty elephants to show our children," we read. "What will the world be like without the Parakaná?"

As we walked down the sloping red dirt street back into town, we met Keila. We had planned to visit Curucu one last time that afternoon and to take Keila with us. She told us to meet her at two, by the boat belonging to her friend Simão, the *Sousa*. She had a surprise for us.

We waited by the *Sousa*, a larger boat than Braulio's *Gigante*. Then, coming over the gentle slope of the sand dunes, we saw Keila, her blonding, curly hair flowing beneath a wide-brimmed straw hat—and behind her, eight of her friends, carrying bags and baskets. They had come, she explained, to perform for us the Dance of the Boto.

They were beautiful, young, lithe women and men: Edson, Nilson, Carisson, Trenato, and Edany, Elailene, Franides, Diolene, the boys wearing baseball caps and muscle shirts, the girls in little black bikinis and pareos. In their bags, they carried their costumes and unlit torches and props. They had brought a cooler full of snacks and drinks.

The *Sousa*, gap-toothed like its owner, spewed smoke and fumes and noise as we chugged to Curucu; Simão bailed water with the dried husk of a gourd. The boys sat in front of the boat with Simão and Keila, while the girls giggled in the back.

Keila's arm shot out as we approached the point. "Boto!" It was Valentino. As the dancers debarked, I swam out to him, and soon I saw they were all there, as if to bid us farewell: David and Gary, Mary and Jesus, Valentino and Pink Lips, Vera and two tucuxis. In the first thirty-five minutes, I counted twenty-six sightings. David and Gary swam very near, pulling their heads from the water to look into my face. Twice I was enveloped in clouds of sizzling bubbles they sent up to embrace my skin. Three of them leapt in a starburst. David leapt high in the air, not ten feet from me; a tucuxi flipped his tail. Valentino came near and rolled over, showing his pink belly.

I did not realize that on the sandy banks of the river Keila and the dancers, Simão and his little daughter, were all watching in astonishment. When I came out of the water, the strongest-looking boy, with two earrings in one ear, asked me in amazement, *"Voce não tem emedo?"* ("You are not afraid?")

"Não—eu gosto!" I replied. ("No, I like it!")

Twice more I entered the water and swam with the botos. I recorded forty-nine sightings in forty-five minutes—and Dianne, on shore, saw many more surfacings, for as usual, without my knowing it, they had surrounded me. And then, it seemed, they moved off.

It was sunset, and a huge fire smoked in the western sky. The dancers were waiting for us on the other side of the point. Keila had set up a little grove of driftwood as props. "My mother doesn't want us to cut trees from the forest," she explained. The young dancers had consumed their drinks and snacks and carefully stowed their garbage away in the cooler. Already, they had donned their costumes. The girls, serious as virgins, wore simple white gowns. They stood facing the river.

The boys, wearing white pants and shirts and hats, began a confident, sinuous saunter toward them, as if swimming. In unison, they began to sing:

> *"Quando o boto virou gente*
> *para dançar no puscirum*
> *trouxe arco, trouxe flecha*
> *e até muiraquitã, e*
> *dançou a noite inteira com*
> *a bela cunhantã."*

> "When the boto transformed to a man
> to dance at the village ball,
> bringing the lucky bow and arrow,
> and carrying the frog-charm for luck and passion,
> he danced all night with the beautiful peasant maiden."

They danced on the beach like lovers, each couple a unit, tenderly cupping the face or touching the hair of the partner, staring into each other's eyes. "They were blind for one another," Necca had said; in this moment, each partner seems to glow with passion. With the sun at

The Dance of the Dolphin: the beautiful girl meets her
dolphin lover at the village ball.

their backs, we could see the shadows of the girls' little black bikinis beneath their white dresses, all innocence and new desire. The boys laid down their partners in the sand and embraced them, singing:

*"Um grande mistério na
roça se fax, fugiu cuhanta
com um belo rapaz . . ."*

"The mystery throughout the land
being the disappearance of the village beauty
with the handsome man . . ."

The sun was melting a golden river on the water, backlighting the dancers. Their features were now indistinguishable. No longer were they these boys and girls; by the magic of their story, they were transformed. Now their song seemed to merge with the voices of the spirits: the Mother of the Vine and the Mother of the Lake, the angels of Music and Opera, the ghosts of divas and murdered Indians and Mario's little son. They called out to us to remember. They sang with the voices of the scientists: Vera and Gary, Miriam and Andrea and Peter, inciting us to probe and to wonder. They sang with the voices of the visionaries: those who create and sustain reserves for both people and wildlife, like Tamshiyacu-Tahuayo and Mamirauá, urging us, even in this era of need and of greed, to hope. They sang with the voices of Violetta and Alfredo, the impossible lovers of *La Traviata*. And they sang in the voices of Moises and Ricardo, Don Jorge and João Pena, reminding us that, here in the Amazon, the most preposterous of impossibilities can come true.

In this way, Keila's young friends became the story, as ancient, as perfect, as distant, and as present as the sun on the water. And perhaps the dolphins knew this, too. When we looked toward the water, we saw the botos had returned. They swam low and slowly, blowing softly, their eyes above the water, watching the drama on the beach.

"E o boto ligeiro na roça fugiu.
Desejando a cabocla
na beira do rio . . ."

"The boto fled through the fields,
desiring the maiden
by the river's edge . . ."

The boys sprang to their feet, and with amazing speed and suppleness, leapt into the river and porpoised through the waves—exactly like dolphins.

We are far from the Manaus opera house and the Meeting of the Waters. Now the air is thick with smoke instead of rain. Yet the journey ends as it had begun beneath the fabulous ceiling frescoes of the Teatro Amazonas: again, I am flanked by Dance and Tragedy.

Out of the water, the dolphin-men emerge. Joyously, each joins his lover, reenacting the promises by which we know the fullness of the world. The botos swim, the dancers dance. But in the western sky, the Amazon is burning.

SELECTED BIBLIOGRAPHY

Following is a list of some of the books and articles most helpful to my research, and some books on related topics that readers might particularly enjoy:

Explorers' and Travelers' Accounts

Bates, Henry Walter. *The Naturalist on the River Amazons.* New York: Dover Publications, 1975.

Cousteau, Jacques-Yves, and Mose Richards. *Jacques Cousteau's Amazon Journey.* New York: Harry N. Abrams, 1984.

Kane, Joe. *Running the Amazon.* New York: Vintage Books, 1990.

Kelly, Brian, and Mark London. *Amazon.* New York: Holt, Rinehart & Winston, 1983.

O'Hanlon, Redmond. *In Trouble Again: A Journey Between the Orinoco and the Amazon.* New York: Vintage Press, 1990.

Rambali, Paul. *In the Cities and Jungles of Brazil.* New York: Henry Holt & Co., 1993.

Shoumatoff, Alex. *The Rivers Amazon.* San Francisco: Sierra Club Books, 1986.

Dolphins and Whales

Bateman, Graham, ed. *All the World's Animals: Sea Mammals.* New York: Torstar Books, 1985.

Cousteau, Jacques-Yves, and Philippe Diole. *Dolphins.* New York: A & W Publishers, 1975.

Kendall, Sarita, Fernando Trujillo, and Sandra Beltran. *Dolphins of the Amazon and Orinoco.* Bogotá, Col.: Fundación Omacha, 1995.

Leatherwood, Stephen L., and Randall Reaves. *Sierra Club Handbook of Whales and Dolphins.* San Francisco: Sierra Club Books, 1993.

Lilly, John C. *Lilly on Dolphins: Humans of the Sea.* Garden City, N.Y.: Anchor Books, 1975.

Perrin, W. F., et al. *Biology and Conservation of the River Dolphins.* Gland, Switz.: IUCN/World Conservation Union, 1989.

Pilleri, Giorgio. *Secrets of the Blind Dolphins.* Karachi, Pak.: Sind Wildlife Management Board, 1980.

Ridgway, S. H., and R. J. Harrison, eds. *Handbook of Marine Mammals.* London: Academic Press, 1989.

Thewissen, J. G. M., ed. *The Emergence of Whales: Evolutionary Patterns in the Origin of Cetacea.* New York: Plenum Press, 1998.

Natural History of the Amazon

Bernardino, Francisco Ritta. *Amazonian Emotions.* Manaus, Braz.: Amazon Multimedia Stock, 1996.

Duellman, William E. *The Biology of an Equatorial Herpetofauna in Amazonian Ecuador.* Lawrence: University of Kansas Museum of Natural History, 1978.

Emmons, Louise H. *Neotropical Rainforest Mammals.* Chicago: University of Chicago Press, 1990.

Forsyth, Adrian, and Ken Miyata. *Tropical Nature: Life and Death in the Rain Forests of Central and South America.* New York: Charles Scribner's Sons, 1984.

Goulding, Michael. *Amazon: The Flooded Forest.* New York: Sterling Publishing, 1990.

Hilty, S. L., and W. L. Brown. *A Guide to the Birds of Colombia.* Princeton, N.J.: Princeton University Press, 1986.

Holldobler, Bert, and E. O. Wilson. *The Ants.* Cambridge, Mass.: Belknap Press of Harvard University Press, 1990.

Kricher, John C. *A Neotropical Companion*. Princeton, N.J.: Princeton University Press, 1989.

Moffett, Mark W. *The High Frontier: Exploring the Tropical Rainforest Canopy*. Cambridge, Mass.: Harvard University Press, 1993.

Walker, Ernest. *Walker's Mammals of the World*, 4th ed. Vols. I and II. Edited by John Paradiso. Baltimore: Johns Hopkins University Press, 1983.

Wallace, A. R. *Natural Selection and Tropical Nature*. London: Macmillan Publishers, 1895.

Wilson, E. O. *The Diversity of Life*. Cambridge, Mass.: Harvard University Press, 1992.

———. *Biophilia*. Cambridge, Mass.: Harvard University Press, 1984.

Mythology, Legends, and Stories

Abram, David. *The Spell of the Sensuous: Perception and Language in a More-Than-Human World*. New York: Vintage Books, 1996.

Beaver, Millie de Sangama, and Paul Beaver. *Tales of the Peruvian Amazon*. Largo, Fla.: AE Publications, 1989.

Roe, Peter G. *The Cosmic Zygote: Cosmology in the Amazon Basin*. New Brunswick, N.J.: Rutgers University Press, 1982.

Slater, Candace. *Dance of the Dolphin: Transformation and Disenchantment in the Amazonian Imagination*. Chicago: University of Chicago Press, 1994.

Smith, Nigel J. H. *The Enchanted Amazon Rain Forest: Stories from a Vanishing World*. Gainesville, Fla.: University Press of Florida, 1996.

Amazonian People, History, and the Future

Collier, Richard. *The River That God Forgot*. New York: E. P. Dutton, 1968.

Davis, Wade. *One River: Science, Adventure, and Hallucinogenics in the Amazon Basin*. New York: Simon & Schuster, 1996.

Descola, Philippe. *The Spears of Twilight: Life and Death in the Amazon Jungle*. New York: New Press, 1996.

Engl, Lieselotte, and Theo Engl. *Twilight of Ancient Peru.* New York: McGraw-Hill, 1969.

Goulding, Michael, Dennis J. Mahar, and Nigel J. H. Smith. *Floods of Fortune: Ecology and Economy Along the Amazon.* New York: Columbia University Press, 1996.

Instituto del Tercer Mundo. *The World Guide, 1997–98.* Oxford, U.K.: Internationalist Publications, 1997.

Kane, Joe. *Savages.* New York: Alfred A. Knopf, 1995.

Page, Joseph. *The Brazilians.* Reading, Mass.: Addison-Wesley, 1995.

Plotkin, Mark. *Tales of a Shaman's Apprentice: An Ethnobotanist Searches for New Medicines in the Amazon Rain Forest.* New York: Penguin Books, 1994.

Smith, Nigel, et al. *Amazonia: Resiliency and Dynamism of the Land and Its People.* Tokyo: United Nations University Press, 1995.

For Advice on Vulture Brains and Other Remedies

Werner, David, with Carol Thurman and Jane Mazwell. *Where There Is No Doctor: A Village Health Care Handbook.* Palo Alto, Calif.: Hesperian Foundation, 1992.

Some Scientific Publications

Azevedo, Aline, et al. *Mamirauá: Management Plan.* Brasilia: Sociedade Civil Mamirauá, National Council for Scientific and Technological Development (CNPq), Environmental Protection Institute of the State of Amazonas (IPAAM), 1996.

Berta, Annalisa. "What Is a Whale?" *Science,* 14 January 1994, pp. 180–181.

Best, Robin C., and Vera M. F. da Silva. *"Inia geoffrensis." Mammalian Species* no. 426 (1993): 1–8.

———. *Preliminary Analysis of Reproductive Parameters of the Boutu, Inia geoffrensis, and the Tucuxi, Sotalia fluviatalis, in the Amazon River System.* Report of the International Whaling Commission, Special Issue UE 6, 1989.

Caldwell, Melba C., David K. Caldwell, and William E. Evans. *Sounds and Behavior of Captive Amazonian Freshwater Dolphins,* Inia geoffrensis. Los Angeles:

Los Angeles Museum of Natural History, Contributions in Science No. 108, 1966.

da Silva, Vera M. F., and Robin C. Best. "Freshwater Dolphin/Fisheries Interaction in the Central Amazon." *Amazoniana* XIV, nos. 1/2 (1996): 165–175.

———. "*Sotalia fluviatalis.*" *Mammalian Species* no. 527 (1996): 1–7.

Gewalt, Wolfgang. "Orinoco Freshwater Dolphins Using Self-Produced Bubble Rings as Toys." *Aquatic Mammals* 15, no. 2 (1989): 73–79.

Gingerich, Philip D. "Origin of Whales in Epicontinental Remnant Seas: New Evidence from the Early Eocene of Pakistan." *Science,* 22 April 1983, pp. 403–406.

Pilleri, G., M. Gihr, and C. Kraus. "The Sonar Field of *Inia geoffrensis.*" *Investigations on Cetacea* 10 (1979): 157–176.

Sylvestre, Jean-Pierre. "Some Observations on Behavior of Two Orinoco Dolphins in Captivity at Duisburg Zoo." *Aquatic Mammals* 11, no. 2 (1985): 58–65.

Thewissen, J. G. M., and S. T. Hussain. "Origin of Underwater Hearing in Whales." *Nature,* 4 February 1992, pp. 444–445.

Thewissen, J. G. M., S. T. Hussain, and M. Arif. "Fossil Evidence for the Origin of Aquatic Locomotion in Archaeocete Whales." *Science,* 14 January 1994, pp. 210–212.

TO CONSERVE THE PINK RIVER DOLPHIN AND ITS HABITAT AND TO VISIT THE DOLPHINS' WORLD

The Rainforest Conservation Fund supports Peru's Tamshiyacu-Tahuayo Community Reserve, where much of the narrative in this book takes place. Members receive a quarterly newsletter and a yearly invitation to visit the reserve with members of its Board of Directors, people whose insights and dedication enliven these pages. For more information, write:

Rainforest Conservation Fund
2038 North Clark Street, Suite 233
Chicago, IL 60614
E-mail: rcf@interaccess.com
Visit RCF's home page at: www.rainforestconservation.org

Much of the early dolphin observations recorded in this book took place at Amazonia Expeditions' lodge near the Tamshiyacu-Tahuayo Community Reserve. The company runs a number of expeditions, guided by an ecologically and culturally sensitive philosophy and an all-Peruvian staff, some of whom you have met in these pages. The company supports RCF with generous, regular donations. To arrange a trip, you may contact:

Amazonia Expeditions
18500 Gulf Boulevard, No. 201
Indian Shores, FL 33785
Tel: 800-262-9669
E-mail: Paul.Beaver@gte.net

The Amazon Conservation Team, founded by ethnobotanist Dr. Mark Plotkin, works to preserve local knowledge of plants, and promotes shamanic study in the Amazon to conserve traditional knowledge and natural habitats. You can become a member by writing to:

> Amazon Conservation Team
> 2330 Wilson Boulevard
> Arlington, VA 22201
> You can visit its Web site at: www.ethnobotany.org

There are now limited opportunities for tourists to visit Mamiraué, Brazil's first sustainable development reserve and site of the Brazilian dolphin and manatee studies featured in this book. Construction of a lodge for visitors was underway as this book went to press. Proceeds from tourism benefit Projecto Mamiraué. To arrange to visit the reserve, contact:

> ricardof@pop-tefe.rnp.br
> Mamiraué also has a home page:
> www.cnpq.br/mamiraua/mamiraua.htm

Earthwatch, a nonprofit organization that pairs interested laymen with field scientists needing research assistance, offers several projects in the Amazon for paying volunteers. A new project explores the ecology of pink and gray river dolphins in Peru's Pacaya-Samiria reserve. For more information, contact:

> Earthwatch Institute
> 680 Mount Auburn Street
> P.O. Box 9104
> Watertown, MA 02471
> Tel: 800-776-0188
> E-mail: info@earthwatch.org
> Web site: www.earthwatch.org

ACKNOWLEDGMENTS

A large number of kind and committed people—scientists, fishermen, conservationists, writers, shamans, editors, fellow travelers, and friends—contributed importantly to this book. Many of them appear in these pages; others inform the text more quietly, and I would like to thank them here.

The idea for this book was born at the Eleventh Biennial Conference on the Biology of Marine Mammals in Orlando, Florida, in December 1995. I am particularly grateful to cetacean experts Thomas Henningsen, Randall Reeves, Brian D. Smith, Fernando Trujillo, and the late Stephen Leatherwood for their generous advice, insights, and information. There I also met Vera da Silva, who later graciously shared her limited time, copious knowledge, and warm friendship during visits to Manaus.

Throughout the four research expeditions for this book, I was blessed again and again with patient and knowledgeable guides to the geographical and spiritual landscape of the Amazonian communities I visited. I am indebted to Moises Chavez, Jorge Soplin, and Ricardo Pipa for their help in Peru. For piloting us safely through lightning storms at the Meeting of the Waters, I thank Nildon Athaide. Miriam Marmontel, Andrea Piris, and Ronis da Silveira guided us in Mamirauá (and also kept Dianne and me from falling out of trees and into caiman jaws). At Alter do Chão, I am grateful to Maria do Socorro D'antona Machado for the succor of her lovely *pousada;* to Braulio for our daily trips on the *Gigante do Tapajós;* to Gilberto Pimentel for the voyage to Rio Arapiuns on the sturdy *Boanares;* and especially to Ludinelda Marino Gonçalvez and Keila Marinho for their stories and dance.

I am indebted to those who translated for me the eloquent words of the local people. For sharing their language skills, impressive local knowledge, and their fine companionship, I thank Isabelle Druant and Jim Penn. For their sensitive translation of the words to the dance of the dolphin, I thank Lilla, Kate, and Jane Cabot and Heather Cumming.

For reading the manuscript and averting numerous errors (in Spanish, Portuguese, Quechua, English, and Latin), I thank: Dr. Paul Beaver, Selinda Chiquoine, Dr. Vera da Silva, Dr. Gary Galbreath, Greg Neise, Dr. Mark Plotkin, Steve Nordlinger, Katy Payne, Dianne Taylor-Snow, and Elizabeth Marshall Thomas. Any errors that remain are my responsibility.

In addition, I acknowledge the special help of the following individuals and institutions:

Dr. Gary Galbreath for taking me with him on his time-travels.

Howard Mansfield for his unending patience, and for being the best editor I have ever had.

Denise Roy at Simon & Schuster and my agent, Sarah Jane Freymann, for their excellent suggestions and enthusiastic support for this project.

Elizabeth Marshall Thomas for her example.

Dianne Taylor-Snow for laughing at inappropriate moments and for the beautiful photographs in this book.

Thanks to:

Anya Antipora; Dr. Márcio Ayres; Andrew Cleve; Ray and Beth Cote; Fred and Rachel Dulin; Jim and Ruth Ewing; Rudy Flores; Dr. Richard Frechette and Monadnock Family Care; Dr. Michael Goulding; Dr. Jon Green and Dr. Joy Schochet; Dr. Peter Henderson; Eliane Ritta Honorato; Mario Huanaquiri and his family; Roxanne Kremer; Dr. William Langbauer; Nelson Lobato; Paul Marks; Dave Meyer; Lee Morgan; David Olive; Tina, Nilda, and Bruna Perilho; Vladimir Pistalo; Felicio Pontes, Jr.; Jim Rowan; Juan Salas; Paul Sterry; Augusto Teran; Dr. John Thornjarbson; and Gretchen Vogel.

And also:

Amazonia Expeditions

Chuckles and the staff of the Pittsburgh Zoo

Instituto Nacional de Pesquisas da Amazonia

The Division of Mammals at the Field Museum of Natural History

The libraries and staffs of Antioch/New England Graduate School and the Harvard Museum of Comparative Zoology

Projecto Mamirauá

Rainforest Conservation Fund